Mathematics the Write Way:

Activities for Every Elementary Classroom

Marilyn S. Neil

EYE ON EDUCATION
P.O. BOX 3113
PRINCETON, NJ 08543
(609) 395–0005
(609) 395–1180 fax

Library of Congress Cataloging-in-Publication Data

Neil, Marilyn S., 1937-
Mathematics the write way : activities for every elementary classroom / Marilyn S. Neil.
p. cm.
Includes bibliographical references.
ISBN 1-883001-19-6 (pbk.)
1. Mathematics--Study and teaching (Elementary) 2. Technical writing. I. Title.
QA135.5.N285 1996
372.7'044--dc20
96-3313
CIP

10 9 8 7 6 5 4 3 2 1

Editorial and production services provided by Richard H. Adin Freelance Editorial Services, 9 Orchard Drive, Gardiner, NY 12525 (914-883-5884)

Published by Eye On Education:

Block Scheduling: A Catalyst for Change in High Schools
by Robert Lynn Canady and Michael D. Rettig

Teaching in the Block
edited by Robert Lynn Canady and Michael D. Rettig

Educational Technology: Best Practices from America's Schools
by William C. Bozeman and Donna J. Baumbach

The Educator's Brief Guide to Computers in the Schools
by Eugene F. Provenzano

Handbook of Educational Terms and Applications
by Arthur K. Ellis and Jeffrey T. Fouts

Research on Educational Innovations
by Arthur K. Ellis and Jeffrey T. Fouts

Research on School Restructuring
by Arthur K. Ellis and Jeffrey T. Fouts

Hands-on Leadership Tools for Principals
by Ray Calabrese, Gary Short, and Sally Zepeda

The Principal's Edge
by Jack McCall

The Administrator's Guide to School-Community Relations
by George E. Pawlas

Leadership: A Practical Role for Principals
by Gary M. Crow, L. Joseph Matthews, and Lloyd E. McCleary

Organizational Oversight:
Planning and Scheduling for Effectiveness
by David A. Erlandson, Peggy L. Stark, and Sharon M. Ward

Motivating Others: Creating the Conditions
by David P. Thompson

The School Portfolio:
A Comprehensive Framework for School Improvement
by Victoria L. Bernhardt

School-to-Work
by Arnold H. Packer and Marion W. Pines

Innovations in Parent and Family Involvement
by William Rioux and Nancy Berla

The Performance Assessment Handbook
Volume 1: Portfolios and Socratic Seminars
by Bil Johnson

The Performance Assessment Handbook
Volume 2: Performances and Exhibitions
by Bil Johnson

Bringing the NCTM Standards to Life
by Lisa B. Owen and Charles E. Lamb

Mathematics the Write Way
by Marilyn S. Neil

Transforming Education Through Total Quality Management: A Practitioner's Guide
by Franklin P. Schargel

Quality and Education: Critical Linkages
by Betty L. McCormick

The Educator's Guide to Implementing Outcomes
by William J. Smith

Schools for All Learners: Beyond the Bell Curve
by Renfro C. Manning

TABLE OF CONTENTS

ABOUT THE AUTHOR

Marilyn S. Neil, Ph.D., is currently Professor of Education at West Georgia College. She first developed her expertise in writing and mathematics as an elementary school teacher. She has written numerous articles and books about mathematics as communication, critical thinking in mathematics, and the link between mathematics and children's literature.

INTRODUCTION

ABOUT THIS BOOK

The teaching of mathematics has been based on the premise that mathematics is a teacher-centered activity where the teacher lectures, demonstrates, and illustrates procedures, while the student sits, listens, asks occasional questions, but mostly just memorizes facts, rules, formulas, and steps. Today there is much emphasis on establishing a more student-centered classroom where the students take an active role in their own learning. Writing in content areas of the curriculum has received much attention as a means to actively involve students in their learning and can serve as a means to empower students as learners.

The purpose of *Mathematics the Write Way: Activities for Every Elementary Classroom* is to present various kinds of writing activities that are appropriate for implementation with elementary students. Writing is an excellent way to enhance students' understandings in mathematics while actively involving them in the learning process.

Teachers of elementary mathematics students who want to make writing a part of their instruction and teachers who wish to implement the 1989 NCTM *Curriculum and Evaluation Standards for School Mathematics* and/or the 1991 *Professional Standards for Teaching Mathematics* will find many suggestions for doing so in *Mathematics the Write Way*. Specific writing activities and various kinds of writing that can be used in mathematics instruction are presented to help increase the mathematical understanding of their students. Various writing activities will help to improve the attitudes of students toward mathematics.

The first chapter of *Mathematics the Write Way* presents a rationale for writing in mathematics. The second chapter presents an overview of the various kinds of writing in mathematics appropriate for elementary students and gives samples of each kind of writing. A description of a learning environment appropriate for writing in mathematics is discussed in the third chapter along with recommendations for creating a classroom conducive to writing in mathematics. The roles of the teacher and the student are also addressed.

In the next chapters, the four common 1989 curriculum standards for teaching mathematics are discussed: Mathematics as Problem Solving, Mathematics as Reasoning, Mathematics as Communication, and Mathematical Connections. You will find writing samples in mathematics which illustrate each standard. Each chapter offers suggestions as to how writing in mathematics can be used to foster the recommendations of the standards.

The final chapter of *Mathematics the Write Way* presents suggestions for ways writing in mathematics can be used to assess the learning of mathematics of elementary students. Appropriate writing samples are also included.

How Samples Were Gathered

Writing in mathematics has been an interest of mine since I started teaching, and I have been collecting samples of writing in mathematics from elementary students for many years. Some of the samples shared in this book were collected from students in classes in which I worked with teachers to help them begin using writing activities in their mathematics instruction. Other samples were collected from students enrolled in summer enrichment programs. Still others samples were supplied by elementary teachers enrolled in graduate mathematics classes I have taught.

The samples were collected from students with diverse cultural backgrounds. Contributors include African American, Hispanic, and Japanese students, and lower and middle class Anglo-American students. Samples were supplied by students at various levels of academic achievement in mathematics — students enrolled in regular mathematics classes, students enrolled in gifted classes, and students enrolled in Title 1 mathematics classes. Due to space considerations, I have included only a small number of writing samples. It has been quite difficult to make selections, but I am hopeful that readers will be able to enhance their own teaching by studying this book.

1

Why Write in Mathematics?

A sixth grade Title 1 student was asked by the teacher to write her opinion about math. The student wrote, "Math is fun. It is my best subject. What I like most is playing games and other things. . . . Right now we are dividing with decimals and it is easy. I have trouble sometimes putting decimals in the right place." Writing is a key communication skill and should be viewed as an integral part of the mathematics curriculum. (NCTM, 1989, p. 28) This is demonstrated when the student described how she feels about mathematics (Fig. 1.1) .

Writing has many benefits because it is a mode of language that particularly lends itself to the acquisition of new knowledge (Emig, 1977). Writing is a means of knowing what one thinks and allows for the discovery of one's own ideas (Bagley and Gallenberger, 1992). Mumme and Shepherd (1990), assert that when students are asked to write about their thinking, we are telling them we value what they have to say and we communicate our confidence in their ability to think.

The *Curriculum and Evaluation Standards for School Mathematics* (NCTM, 1989) assert that writing about mathematics, such as describing how a problem was solved, helps students clarify their meaning and develop deeper understanding (p. 26). The *Standards* document suggests that communication in mathematics is vital in the process of concept development and in the ability of students to analyze and use mathematical reasoning. Writing is an important means of communicating. Mathematics instruction should involve student-centered communication because communication helps students enhance their understanding of mathematical concepts. Communication empowers students as learners, promotes a more comfortable, less stressful, discourse

FIGURE 1.1. FEELINGS ABOUT MATHEMATICS

Math is fun. It is
my best subject what I like
most is playing games
and other things I have a
nice teacher. She is really
nice. I am lucky to have
a teacher like her she lets
us play games on friday
if we come in quit and seat
down. Right now we
are divided with decimals
and it is easy. I have
trouble sometimes putting
decimals in the right place
Heres and example of what:
I'm doing in math.

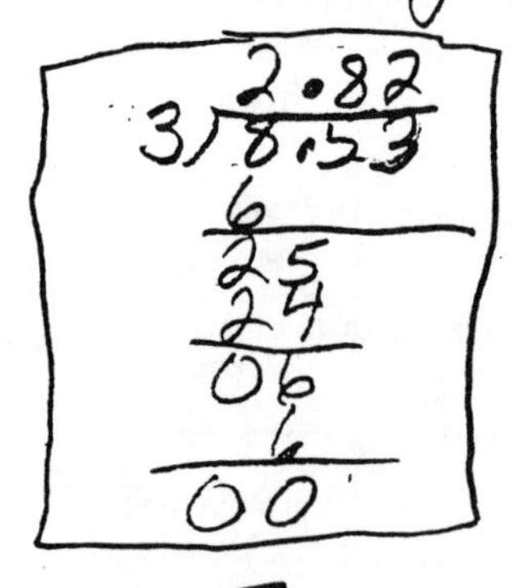

Example

for learning and assists teachers in gaining insights into students' thinking (Mumme and Shepherd, 1990).

Writing is a means through which students communicate their understandings of mathematics and its application (Miller, 1991). Writing represents a unique mode of learning, not merely valuable, not merely special, but unique (Emig, 1977).

- **Writing provides opportunities for students to construct their own knowledge and understanding of mathematical concepts and procedures** (NCTM, 1991).

In constructing their own knowledge, students learn to reason mathematically, and to conjecture, invent, and solve problems (Winograd and Higgins, 1995). In writing, learners can actively construct their own meaning. As they think about experiences, both in and out of school, learners are able to construct meaning by integrating new ideas with those meanings already acquired.

- **Writing is a means for fitting new events into existing structures.**

Heibert (1990) identified three cognitive processes that contribute to the construction of mathematical knowledge. During Heibert's **representation** and **reflection** phases of learning, learners can share what they have done in writing in order to clarify and expand meaning. Payne (1990) stated that Heibert's **construction** of conceptual knowledge requires the active cognitive effort of the learner. Writing is an active process, therefore is appropriate for learners in constructing meaning.

- **Writing provides a means for students to explore their thoughts as they communicate their ideas** (Stix, 1994).

When students communicate their thoughts in writing, they find out what they think. Writing helps heighten and refine thinking (Olson, 1984). Students reflect on, and in turn clarify, their thinking. Writing is a thinking tool that forces a slowdown in thought processes, frees the brain to play around with ideas, make new discoveries and to more fully integrate new knowledge (Gere, 1985). Writing enhances cognitive development and (the writer) utilizes higher-order thinking (Thornton, 1991).

- **Writing allows students to clarify, record, and demonstrate their learning processes and outcomes** (Richards, 1990).

Writing increases recall and understanding of information and reinforces the learning of specific content. When students write they are able to organize; their writing provides an opportunity for them to reframe newly acquired knowledge in their own words (Wilde, 1991). Through writing, students are able to think about the mathematical concepts they are writing about and to focus on and internalize important concepts. Writing enables students to personalize mathematical concepts.

In her explanation of a fraction (Fig. 1.2), the student begins the work of defining a fraction. She makes illustrations to further explain her meaning of a fraction.

- **Writing permits students to apply new concepts.**

Students may apply newly acquired mathematics concepts in many ways. They may write and publish a manual that explains how to solve problems in mathematics make constructions using geometry or measurement concepts they have acquired. Students may write original story problems that show their discoveries and reveal their understandings of mathematical concepts. When students create original stories about familiar situations, they learn to value problem-solving skills. Writing is a vehicle for applying thinking in other areas of the curriculum.

A third grade student read *Alexander Who Used to Be Rich Last Sunday* by Judith Viorst. This is a story in which Alexander is given $1.00. Instead of saving the money, he spends, squanders, or loses his money until he has none.

When the student finished reading the book, his teacher asked him to write a parallel, or "copycat," story of the book. The pattern modeled in *Alexander* furnishes a ready-made framework within which the student could develop a personal context using the mathematical concept illustrated (Lewis, Long, and Mackay, 1993).

FIGURE 1.2. EXPLAINING A FRACTION

What Is A Fraction? 4-25-94

A fraction is a way of dividing. Its almost just like multiplying. Muttiplying is just adding again and again and fractions is just dividing. Lets say you made a pizza with 8 pieces. Including you, four people are going to eat it. You need to to know how many pieces each person is going to have. This is the fraction. (8÷4=2.) Each person has two pieces.

Examples

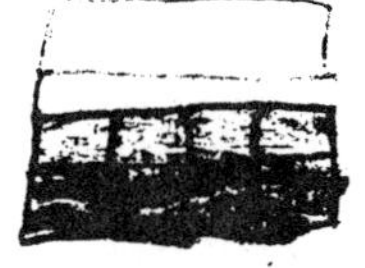

$\frac{1}{2}$ is yellow.
$\frac{1}{2}$ is purple

$\frac{2}{3}$ is yellow
$\frac{1}{3}$ is purple

On the first page of his story (Fig. 1.3), the student wrote, "Last Sunday, I used to be rich, but now I'm broke. All I have is a half eaten dounut. Last Sunday my grandmamma and grandpapa came over and gave me ten whole dollars! My mouth watered with greed." He continued, "I said I was gonna save that money but I went to Baskin Robbins and spent $1.75 on a vinnila ice cream cone. There goes $1.75." Beside the illustration of a large ice cream cone, the student wrote a subtraction algorithm in which he correctly subtracted $1.75 from $10.00.

The third-grade student continued writing and illustrating his parallel story. On Tuesday, he broke a lamp and had to pay $2.95. His story included an illustration of the lamp falling from a table after it was accidentally hit by a basketball. Note that he correctly subtracted $2.95 from $8.25 and included the algorithm beside the illustration. Then the student creatively ended his story: "I said I was gonna save my money, but on Sunday, I littered on a highway and the cops fined me $300.00." At this point, he only had 55¢ left. He asks, "How am I gonna pay it?" The parallel story written by this third-grader clearly supports the role of writing in mathematics in applying new concepts. His story also illustrates how he explored his thoughts as he communicated his ideas in writing.

- **Writing helps students organize their ideas into a synthesis of ideas that will make sense to them and their readers** (Fortescue, 1994).

As students apply concepts and develop confidence in their own abilities, they come to see mathematics as meaningful and relevant in their everyday lives. When students write about mathematics, they must examine the problem, its elements, and its relationships. Students become familiar with analytical writing as they display a deeper understanding of a concept (Fortescue, 1994). As students become proficient in mathematics, they build their confidence as mathematicians. They can review, reiterate, and deepen understandings of mathematical concepts (Kay and Charles, 1995).

(Text continues on page 13.)

FIGURE 1.3. PARALLEL OR "COPYCAT" STORY (SELECTED PAGES)

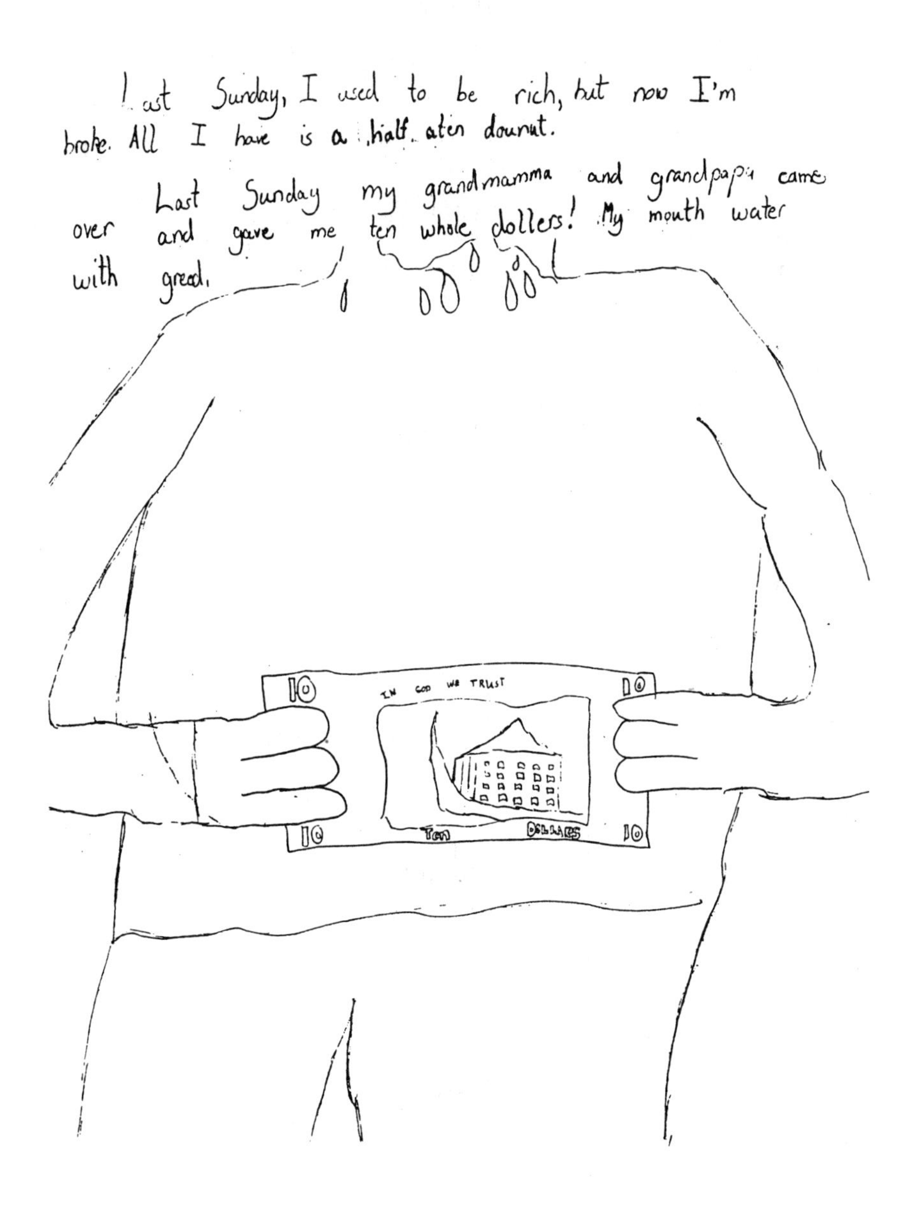

I said I was gonna save that money but I went to Baskin Robbins and spent $1.75 on a vinnila ice cream cone. There goes $1.75

I said I was going to save that money, but on Tuesday, I kinda broke the lamp, and had to pay $2.95. There goes $2.95

I Said I was gonna save my money, but on Sunday, I littered on a highway and the cops fined me $300.00. How am I gonna pay it?

In Figure 1.4 (on the next page), a student explains what was done incorrectly in solving a problem. Her written explanation shows that she understands the standard algorithm which suggests she subtract 5 – 1. Written explanations such as this allows the teacher to evaluate understandings of students.

- **Writing helps students formulate, organize, internalize, and evaluate concepts** (Nahrgang and Petersen, 1986).

When students write informally and in their own words, they reflect on their real-world experiences. These personal writings allow them to organize mathematical concepts and experiences in such a way that the concepts and experiences become their own. In Figure 1.5, the student describes a mathematical experience in which the class made cookies. The student understands the necessity of applying the mathematical concepts of measuring amounts and temperature when making cookies. It appears he has internalized the importance of measurement in an everyday real-world experience.

FIGURE 1.5. APPLYING MEASUREMENT AND TEMPERATURE

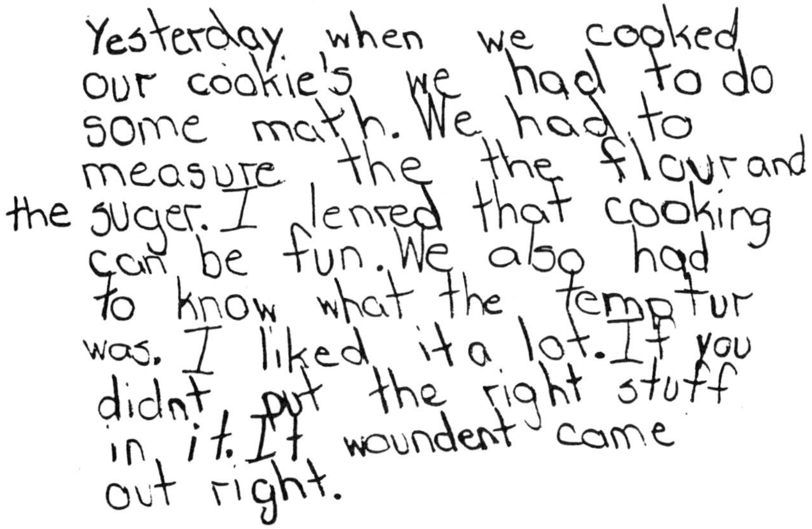

Yesterday when we cooked our cookie's we had to do some math. We had to measure the the flour and the suger. I lenred that cooking can be fun. We also had to know what the temptur was. I liked it a lot. If you didnt put the right stuff in it. It woundent come out right.

FIGURE 1.4. EVALUATING STUDENT UNDERSTANDING

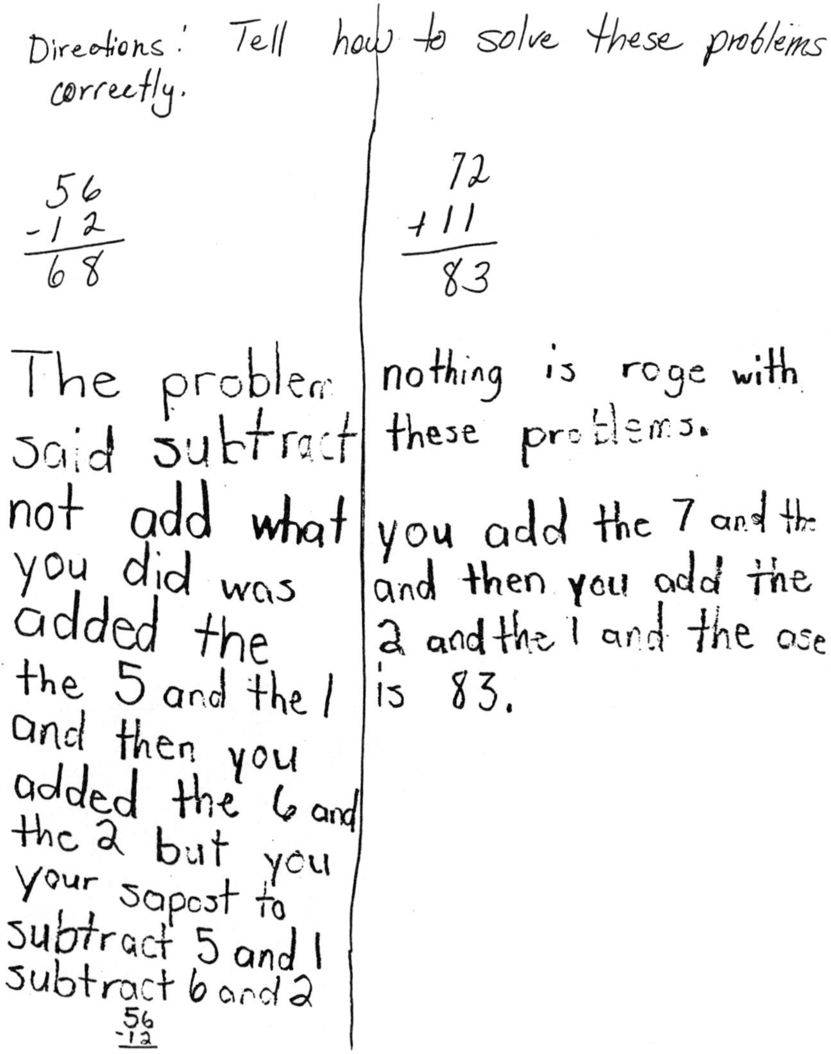

Directions: Tell how to solve these problems correctly.

56 -12 68	72 +11 83
The problem said subtract not add what you did was added the the 5 and the 1 and then you added the 6 and the 2 but you your sapost to subtract 5 and 1 subtract 6 and 2 56 -12	nothing is roge with these problems. you add the 7 and the and then you add the 2 and the 1 and the ase is 83.

HOW WRITING IN MATHEMATICS REFLECTS THE NCTM'S 1989 AND 1991 STANDARDS

Writing in mathematics is one way to implement the 1989 NCTM *Curriculum and Evaluation Standards for School Mathematics*. Four standards are common for the K–4, 5–8, and 9–12 mathematics curriculums. The four common standards are mathematics as problem-solving, mathematics as communication, mathematics as reasoning and mathematical connections.

Mathematics as communication is the more obvious standard for inclusion of writing. This standard calls for the integration of language arts and mathematics as students write in mathematics. When students describe how a problem is solved, they are clarifying their thinking and developing deeper understanding in mathematics.

Writing in mathematics can be an integral part of classroom instruction for implementing the mathematics as problem solving standard. Students can share their thinking and their approaches to problem-solving with other students and with teachers. The sharing of thinking and approaches used in thinking can be implemented through writing in mathematics.

It is important that students have many experiences in creating problems from real-world activities. Writing in mathematics is an excellent means for students to record word problems created after participating in real-world activities. When problem-solving is emphasized in the classroom, students can grow in their ability to communicate mathematically, particularly in writing in mathematics.

Writing in mathematics is also applicable to the mathematics as reasoning standard. In the *Standards* document, it is suggested that classrooms that value reasoning also value communication and problem-solving. Writing and other forms of communication are vehicles for implementing this standard. When students justify their solutions, as suggested in the standards, they can do so by writing in mathematics.

When students compare and contrast concepts and procedures as recommended in the mathematical connections standard, they can do so through their writing. Students can explore meanings, make illustrations, and describe through writing ways mathematics is used in real-life experiences. The study of mathematics should include opportunities for students to use mathematics in other curriculum areas and writing in mathematics is an important vehicle for making mathematical connections.

WRITING IN MATHEMATICS IS SUPPORTED IN THE NCTM'S *PROFESSIONAL STANDARDS FOR TEACHING MATHEMATICS* (1991)

Writing in mathematics supports and is supported by the NCTM's *Professional Standards for Teaching Mathematics* (1991). These standards support the use of writing in mathematics as an instructional technique because writing is a worthwhile task (Stewart and Chance, 1995). In their search of the literature about writing in mathematics, Stewart and Chance identified numerous connections between writing and NCTM's *Professional Standards for Teaching Mathematics* (1991). The connections found in this study have implications for writing in K–6 classrooms.

Standard 1 of the NCTM's *Professional Standards* suggests that the interests, understandings, experiences, and learning styles of students should be the basis for the selection of mathematical tasks for use in instruction. Tasks selected for instruction should engage students' intellect and build a framework for reasoning and problem-solving in mathematics. The writing of journals, learning logs, and original story problems are recommended ways for meeting this standard.

When students write in mathematics, they are intellectually engaged in that they must actively think about what they write. Their mathematical understandings and skills are enhanced when they explain, define, identify relationships, or list steps in journals or learning logs. Students make connections and develop a coherent framework for mathematical ideas as they write in journals or learning logs. When students formulate and solve their own original problems in writing, communication about mathematics is promoted.

Standard 2 of the *Professional Standards* proposes that teachers of mathematics should pose questions and tasks that elicit, engage, and challenge students' thinking, and ask students to clarify and justify their ideas in writing (p. 35). Students can record their ideas and provide a representation of their reasoning. Writing in mathematics in K–6 classrooms allows this standard to be implemented, and writing, in turn, supports this standard.

Writing is supported by Standard 3 of the *Professional Standards.* This standard suggests that the teacher promote classroom discourse by using a variety of tools. Writing is a useful tool for supporting discourse.

Standard 4, "Tools for Enhancing Discourse," can also be met through writing in mathematics. Tools suggested in this Standard include using models, pictures, diagrams, tables, graphs, metaphors, analogies, stories, and written hypotheses, explanations, and arguments in the mathematics classroom (p. 52). Writing in mathematics, whether presenting explanations or describing procedures, is one appropriate way to meet this standard.

The learning environment is addressed in Standard 5 of the *Professional Standards.* It is recommended that the teacher of mathematics create a learning environment that provides a context which encourages the development of mathematical skill by using materials in ways that facilitate students' learning of mathematics. Certainly, when students are free to write, describe, explain, and justify, mathematical skills are enhanced and proficiency in mathematics is fostered. The writing classroom becomes more student-centered in that students work independently, take risks by asking questions in writing, and take control of their learning of mathematics. Writing in mathematics supports and is supported by this standard.

Standard 6 of the *Professional Standards,* "Analysis of Teaching and Learning" (p. 63), suggests that teachers of mathematics gather information about students in order to assess what they are learning. An excellent way of assessing what students are learning in mathematics is through the careful examination of students' writing. When students make recordings in journals, describe procedures in learning logs, or write original story problems, teachers can examine the writings to gather information about what students do and do not understand and, therefore, assess their understanding and plan for further experiences and instruction.

GENERAL SUGGESTIONS FOR WRITING IN MATHEMATICS

- Provide many opportunities for students to write in mathematics, not just during mathematics classes.
- Provide students with writing role models, such as those in good children's literature. Read to and with students, and have students read good literature.
- Provide students with many experiences for gathering information, such as hands-on activities, field trips, books, and magazines to give them backgrounds for writing.
- Provide many opportunities for interaction of students with other students and with the teacher for communicating and sharing ideas, concepts, and writings. These opportunities include working with partners, in small groups, and in whole group discussions.
- Give students many experiences with physical materials (including manipulatives) to provide a background of experiences to promote communication among students and to provide first-hand knowledge for students; both are necessary for writing.
- Have students write and illustrate their own story problems. Student-created stories have meaning for students.
- Have students brainstorm in small group or large group settings and record and classify terms or concepts to be used in their writings. Encourage students to discuss their classifications.
- Present many open-ended and/or probing problems for students to solve.
- Create a learning environment in which all students feel comfortable and where they can successfully do mathematics and write about their experiences.
- Integrate language arts with mathematics. Students can extend their writing skills in meaningful ways as they write in mathematics.
- Use the students' own language in developing mathematical problems for them to solve.

- Expose students to models of the kinds of problems they will write (Winograd, 1993). Models should come from the teacher, other students, and from the formal curriculum.

REFERENCES

Bagley, Theresa, and Gallenberger, Catarina. "Assessing students' dispositions: Using journals to improve students's performance." *The Mathematics Teacher* (November, 1992), pp. 660–663.

Emig, Janet. "Writing as a mode of learning." *College Composition and Communication* (May 1977), pp. 122–128.

Fortescue, Chelsea M. "Using oral and written language to increase understanding of math concepts." *Language Arts* (December, 1994), pp. 576–580.

Gere, Ann Ruggles. *Roots in the Sawdust: Writing to Learn Across the Disciplines.* Urbana, IL: National Council of Teachers of English (1985).

Hiebert, James. "The role of routine procedures in the development of mathematical competence." In: Cooney, Thomas J., and Hirsh, Christian R., editors. *Teaching and Learning Mathematics in the 1990's.* Reston, VA: The National Council of Teachers of Mathematics (1990).

Kay, Cynthia, and Charles, Jim. "Integrating math and writing." *Teaching K–8* (May 1995), pp. 22–23.

Miller, Janet. "Writing to learn mathematics." *The Mathematics Teacher* (October 1991), pp. 516–521.

Mumme, Judith, and Shepherd, Nancy. "Communication in mathematics." *The Arithmetic Teacher* (September, 1990), pp. 18–22.

Nahrgang, Cynthia L., and Petersen, Bruce T. "Using writing to learn mathematics." *The Mathematics Teacher* (September, 1986), pp. 461–465.

National Council of Teachers of Mathematics. *Curriculum and Evaluation Standards for School Mathematic.* Reston, VA: The National Council of Teachers of Mathematics (1989).

National Council of Teachers of Mathematics. *Professional Standards for Teaching Mathematics*. Reston, VA: The National Council of Teachers of Mathematics (1991).

Olson, Carol Booth. "Foster critical thinking skills through writing." *Educational Leadership* (November, 1984), pp. 28–39.

Payne, Joseph N. "New Directions in mathematics education." In: Payne, Joseph, editor. *Mathematics for the Young Child*. Reston, VA: The National Council of Teachers of Mathematics (1990).

Richards, Leah. "Measuring things in words: Language for learning mathematics." *Language Arts* (January 1990), pp. 14–25.

Stix, Andi. "Pic-Jour math: Pictorial journal writing in mathematics." *The Arithmetic Teacher* (January, 1994), pp. 264–269.

Thornton, Carol A. "Think, tell, share—success for students." *The Arithmetic Teacher* (February, 1991), pp. 22–23.

Wilde, Sandra. "Learning to write about mathematics." *The Arithmetic Teacher* (February, 1991), pp. 38–43.

Winograd, Ken, and Higgins, Karen M. "Writing, reading, and talking mathematics: One interdisciplinary possibility." *The Reading Teacher* (December, 1994), pp. 310–317.

2

Types of Writing in Mathematics for K–6 Students

Writing promotes meaningful learning and mathematics can be communicated in writing. Vacca and Vacca (1989) suggested creating occasions for students to write regularly because they believe writing is a powerful strategy for learning in content areas, including mathematics. Many writing strategies are appropriate for K–6 students. In this chapter, four types of writing, expository writing, expressive writing, creative writing, and using the writing process in mathematics, are described in detail and samples of each are shared. Several other strategies appropriate for writing in mathematics are also briefly described and illustrated.

EXPOSITORY WRITING

Expository writing in the form of note taking, copying from the chalkboard or textbook, recording, and summarizing has been used to a great extent in mathematics. The primary purpose of expository writing is to record and/or explain content in mathematics or other content areas. Expository writing includes:

- Describing how a problem has been solved;
- Recording predictions for estimations or problem solutions;
- Writing instructions about how to make a material for mathematics or instructions for playing a game in mathematics;
- Rewriting a story problem from a textbook in the student's own words;
- Describing the thought processes used to arrive at an

incorrect solution (Artz, 1994);

- Writing a self-help sheet for a friend (Cook, 1995);
- Creating a word web where students respond to a selected key word or phrase with knowledge they have about the word (McGehe, 1991);
- Recording questions, observations, lists, how to's, or summaries of what has happened in mathematics class; and
- Writing and publishing a newspaper article or column describing one or more activities in mathematics (McIntosh, 1991).

Expository writing involves the students' thinking and helps make knowledge in mathematics the student's own. Students focus on and record information about mathematics during expository writing. They can later retrieve the information because it has been recorded. These recordings (expository writing) can be kept in a spiral or three-ring notebook, often referred to as a **learning log.**

In Figure 2.1, a student explained how she felt about mathematics and then showed an error she made when subtracting $43.64 from $52.00. She then recorded the correct answer.

Learning logs permit students to pose questions, pursue possible answers, arrive at conclusions, and to review and reflect on understandings, concepts, data, brainstormed lists, and steps related to mathematics.

Students may write simple lists or phrases, simple sentences and paragraphs, or they can compose longer pieces. Expository writing is an effective type of writing to be used in mathematics.

EXPRESSIVE WRITING

One type of writing recommended for mathematics is expressive writing where students express and document their mathematical thoughts (Helton, 1995). In expressive writing, students do the equivalent of thinking out loud, but record their thinking on paper. Gere (1985) stated that expressive writing is a "thinking tool" which forces a slowdown in thought processes allowing students to play around with ideas, make new discoveries, and inte-

FIGURE 2.1. LEARNING LOG—EXPLAINING AN INCORRECT ANSWER

I sorda like math it's ok, but when it gets to something hard I try my best. And sometimes I past, like on the sub traction I did this:

I came up with error with 9.36 → 52.00 − 43.64 = 09.36

And when I tried Again I got it right. Answer → 52.00 − 43.64 = 08.36

grate new knowledge.

Expressive writing is spontaneous and less formal than other types of writing. In expressive writing, students engage in free writing or respond to prompts (impromptu writing). Students may reflect on and evaluate their feelings and opinions about learning mathematics or they may write about the thinking processes they use for solving problems.

Journal writing is a form of expressive writing. This type of writing offers an interactive way for students to communicate what they have learned. It is considered by many to be the most effective method of using writing to teach mathematics (Vacca and Vacca, 1989). In journals, students may free-write or respond to a prompt; they may respond to literature with a mathematics context, or revise a story to include numerical information. Students may complete a guided response sheet or a word web.

As early as kindergarten, students can engage in journal writing. They may begin by drawing a picture representing what they have created or learned while participating in a discussion or manipulative activity in mathematics. Stix (1994) used *Pic-Jour Math* journals with young students. Young students can use pictures, numbers, symbols, and manipulatives to help them better understand concepts in mathematics while making mathematics their own. After writing in "pictorial" journals, students' ability to define concepts and present a clear, logical solution to a given task were enhanced.

Figure 2.2 is an illustration from a picture journal by a kindergarten student whose teacher, Angie Hall, had read *Ten Apples Up On Top* (Le Sieg, 1961). The student's writing was limited, but he was able to clearly illustrate the mathematics concept being explored in the book.

Wendy Wicht began journal writing in mathematics with her third-grade class by first discussing the purpose of writing in journals. She explained that the journals were like diaries and she began by having her students respond to questions or statements called prompts. In journals, Wendy's students explored their attitudes and understandings when they wrote about what they learned in mathematics. Students first responded to the two prompts—"What I like about math" and/or "What I don't like about math" (Fig. 2.3).

FIGURE 2.2. KINDERGARTEN PICTURE JOURNAL

FIGURE 2.3. EXPLORING ATTITUDES ABOUT MATHEMATICS

What I Like about math.

I Like some parts of math
like Measurement and Multiplying
But the tow I like more is
the Calculator and problem
solving. I like math because
my friend Carrie is in hear and
she is nice. I Like math
because it is the last Subject.
I like math because we eat
I like math because we get
a nice teacher wean you go
to the sore you Do matho. math
it so fun!

What I don't like about math

Math is hard sometimes and
I hate divide.

Helton (1995) suggested asking students to respond to opinion questions (also referred to as prompts) such as:

"What did I learn in math today?"
"What did I like about what I learned today?"
"What did I not like about what I learned today?"

When students express their thoughts in journals, the teacher should respond in writing to students. This dialogue becomes a written conversation between student and teacher where students are free to express their own ideas, thoughts, questions, concerns, and reactions about mathematics, and teachers respond to students' entries. Students are then able to share privately their concerns and ideas without threat of evaluation by the teacher because a finished writing product is not expected (Gordon and Macinnis, 1993). A simple response by the teacher is all that is necessary to many entries, as shown in Figure 2.4, where the teacher simply answered the student's question, "Do you like math?"

FIGURE 2.4. TEACHER RESPONSE

Expressive writing also can be free writing where students write daily for 10 to 15 minutes, or even only 1 or 2 days during a week, about anything they choose in mathematics. This free writing encourages the independent discovery of ideas (Gordon and Macinnis, 1993). In free writing activities, students may use disconnected phrases, incomplete sentences, and/or misspelled words (Burton, 1985).

CREATIVE WRITING

Another type of writing appropriate for mathematics is creative writing. Creative writing has traditionally only been done in language arts. Creative writing may include writing original poems about mathematical concepts, writing original stories or mathematics problems, writing letters to mathematicians, or writing original plays (McIntosh, 1991). Winograd (1993) stated that when students are invited to write their own stories and/or problems, the mathematics curriculum becomes more authentic and conceptual.

For students to creatively write in mathematics, they must engage in many authentic experiences. These experiences provide a background for writing. The writing can include experiences such as trips to the supermarket, reading, describing tables and graphs from newspapers or magazines, listening to or reading books and stories containing mathematical content or ideas, writing original or parallel stories, attending sporting events, or having spend-the-night parties and writing about them, including in each script a mathematical context.

Students can write creative stories about familiar situations involving mathematics concepts, such as time. One student's book entitled "Spend-the-Night Party" (Fig. 2.5) is an example of a story about time. This student's teacher stamped a clock with no hands in the upper right-hand corner of several blank pages for students. Each student wrote an original story involving time. In their stories, students described things that happened at various times and they then drew the hands on the clock on each page of their story to show the time the happenings took place.

(Text continues on page 35.)

FIGURE 2.5. CREATIVE STORY INVOLVING THE CONCEPT OF TIME (SELECTED PAGES)

At 6:30 every body
came over to my house.

At 8:30 we had

Cool!

At 1:00 Am we played truth or Dare.

Figure 2.6 shows an original story problem written by a third grade student about the number of legs on sheep. Multiplication was being explored in the student's class and the teacher encouraged students to pose original multiplication story problems. Silverman, Winograd, and Strohauer (1992) support the writing of original mathematics story problems because they believe these problems "replace the dull, predictable, textbook problems." (p. 6).

FIGURE 2.6. ORIGINAL TWO-STEP STORY PROBLEM

There are 6 sheep each sheep
have 4 legs if 2 more sheep
jowned how many legs would there
be in all

Susan Smith, a kindergarten teacher, read *Eating the Alphabet,* by Lois Ehlert, to her class. She discussed with students the letters of the alphabet and the fruits and vegetables in the book. Students identified various fruits from the book and Susan brought apples, pineapple, raspberries, bananas, and grapes to class. Students then selected fruit and made "fruit kabobs" by placing the pieces of fruit in simple **patterns**. Next students drew, colored, and cut out drawings of the fruit kabobs and made patterns using the cutout fruits. Then they pasted the cutout fruit patterns on paper. Susan put the cutouts together and "published" them as a book entitled *Patterned Fruit Kabobs.* The book made by the students was a creative way for them to record their explorations and gave them a way to relive their experience at a later time when they reread their book which Susan had placed in a reading center.

Margaret Innocenti asked her sixth grade students to write creative math stories using "picture cards." She cut pictures from magazines, pasted them on construction paper, and laminated them. She held a group discussion in which students described what they saw in the pictures. They discussed the meaning of a problem and then worked in groups to create original mathematics stories about a selected "picture card." Later, the pictures and story problems were exchanged among groups where problems were evaluated and later solved.

Students can also create poems and plays or playlets that include mathematical content. Hosmer (1986) stated that even young children can write stories before they have developed the skills needed for clear explanation. Clearly, creative writing with mathematical contexts is appropriate for K–6 students.

THE WRITING PROCESS

The 1989 NCTM *Curriculum and Evaluation Standards for School Mathematics* suggests that writing can be used as a strategy to make problem-solving meaningful to students (Ford, 1990). Ford reported findings from the works of Donald Graves and Lucy Calkins. These findings suggest that guiding students through a writing process helps produce better thinkers and, as a result of the writing experience, students become better problem-solvers. Problem-solving and writing are both processes and it is generally accepted that when students experience the writing process in mathematics and use their own language to develop story problems, their problem-solving abilities are enhanced. Ford further suggested that if students move through all the stages of the writing process this becomes a successful strategy for problem-solving. Additionally, students who apply the writing process to their problem-solving frequently develop positive attitudes toward problem-solving.

Most early childhood and elementary teachers are familiar with the five basic stages in the writing process. The stages are:

- Prewriting;
- Writing;
- Conferencing;

- Revising; and
- Publishing.

Donald Graves (1983), an authority on writing, stated that the writing process has many beginning points, including what he refers to as rehearsal. He indicated that the writer must prepare for composing and that the rehearsal may include outlining, reading, conversing, or writing lines (p. 221). This rehearsal stage is often referred to as "prewriting." A stimulus is provided by the teacher (Ford, 1990) and students identify topics, brainstorm ideas, or simply write beginning ideas about a topic. This first stage in the writing process is similar to the first stage in the mathematical problem-solving process where students identify the problem.

In the prewriting stage, students think about what they already know about the situations, ideas, or concepts. The teacher may work with the large group or place students into small groups to brainstorm and examine ideas. The prewriting or brainstorming stage acts as a stimulus for writing. Ford (1990) described a situation in a class of third graders where she used the writing process to help improve students' attitudes toward mathematics and their problem-solving skills. As a part of the prewriting experience, she distributed menus and catalogs and gave students an example of a situation in which she and her sister went out to eat. Students were asked to think about similar events in their own lives.

Graves (1983) refers to the second stage of the writing process as composing. He stated that "composing refers to everything a writer does from the first time words are put on paper until all drafts are completed" (p. 223). Ford followed Grave's suggestions and had students write their own ideas for math problems after looking through the menus and catalogs.

Next students worked in pairs, read what their partners had written, and solved the problems created (conferencing stage). During this stage students discussed if there was a problem with their partner's math problem and each helped their partner create an acceptable math problem. Next was the revising and editing stages where students corrected their own problems. Finally, students published (the last stage) their problems by writing them on index cards to be placed on the mathematics table.

Other activities appropriate for use during the prewriting process include presenting pictures, newspaper ads, tables, charts, or graphs for discussion and examination. Appropriate ways for publication of students' problems or stories include placing problems on posters to be placed in the classroom, making problems into transparencies to be used on the overhead projector or binding student problems into classroom books, perhaps after they have been illustrated (Fennell and Ammon, 1985).

Figure 2.7 shows two original story problems that were written by sixth grade students. These story problems were edited, illustrated, and then published. Ms. Innocenti, the teacher, directed students through the writing process. One student reflected a real situation in his problem, while the other student seemed to create a problem from his imagination. In both cases students reflected their understanding of composing word problems and applied their understanding of multiplication and subtraction. Problem-solving was becoming meaningful to them.

OTHER TYPES OF WRITING

Matz and Leir (1992) described the "stop method" or word-problem playlet, as another way that students can learn to solve complex word problems. Three students wrote a word-problem playlet about buying pets at a pet store (Fig. 2.8). The students decided on a setting, the characters, and an original problem. The students then performed their playlet for the rest of the class. When the characters presented enough information, they asked a question and then said "Freeze." Members of the class then had to use the information presented in the playlet to solve the problem.

Matz and Leir (1992) state that the stop method, role playing, and authorship all come together when students write and perform their own word problem playlets using this technique which is called **stage freeze** (p. 15). Realism is added to the word problems constructed by a group, and students analyze the problem situations by making a "mental script" (Matz and Leir, 1992).

FIGURE 2.7. ILLUSTRATED AND PUBLISHED ORIGINAL STORY PROBLEMS

John bought a stereo from Best Buy a week ago. It cost $3,999.99. A week later Best Buy had a sale. The stereo then cost $3,031.99. How much money would John had saved if he bought the stereo a week later?

There are 21 crabs in the wagon. The boy pulling the wagon has to pull 250 more wagons load of crabs. How many crabs does he have to pull?

FIGURE 2.8. WORD PROBLEM PLAYLET WITH "STAGE FREEZE"

Pet Store

Latoya: "Caroline lets go to the pet store".

Caroline- "Hey that sounds like a fun idea

Latoya "Jame what are you doing here"?

Jame "This is my summer job".

Jame " What would you like to buy Caroline"?

Caroline "I'd like to buy a rabbit".

Jame "A rabbit is $17.27".

Caroline "Great I brought a one hund- dollar bill"!

Jame "Caroline your Change is plus ta is"? Freeze!

Jame "Latoya what would you like to buy"?

Latoya "I want to buy a hampster".

Jame "A hampster costs $5.21".

Latoya "Great I brought a one hundred dollar bill, too"!

Jaine "Latoya your Change is plus tax is Freeze!

McGehe (1991) suggested another form appropriate for writing in mathematics. She described how students use a **word web** as a way to organize and clarify problem-solving strategies. A key word or phrase is written in the middle of a chart or on the chalkboard, and students respond by listing anything they know or feel about the phrase or concept. As responses are recorded by the teacher, she organizes them into categories. Placement of responses are discussed and students suggest where to place responses. Word webs can also be done by individual students and recorded in their journals or logs.

Wood (1992) discussed how **reaction guides** can be used as a strategy to stimulate the thinking of students. She suggested that reaction guides stimulate students' thinking after a unit or lesson and described the reaction guides as consisting of five to eight statements that reflect both concepts and misconceptions about statements such as "The meter is a unit of length in the metric system" and "The only reason we need to study the metric system is because it's in our math textbook" (p. 98). Students work in pairs or small groups to refute or confirm each statement and record their evidence (support) in writing into a reaction guide. These guides serve as a review after instruction about a selected concept or lesson.

Many types of writing in mathematics appropriate for K–6 students were described in this chapter. Later chapters explore each writing type and provide samples of each type of writing.

REFERENCES

Artz, Alice F."Integrating writing and cooperative learning in the mathematics class." *The Mathematics Teacher* (February, 1994), pp. 80–85.

Burton, Grace M. "Writing as a way of knowing in a mathematics education class." *Arithmetic Teacher* (December, 1985), pp. 40–45.

Cook, Jimmie. "Integrating math and writing." *Teaching K–8* (May, 1995), pp. 22–23.

Ehlert, Lois. *Eating the Alphabet: Fruits and Vegetables from A to Z.* New York: The Trumpet Club (1989).

Fennell, Francis, and Ammon, Richard. "Writing techniques for problem-solvers." *Arithmetic Teacher* (September, 1985), pp. 24–25.

Ford, Margaret I. "The writing process: A strategy for problem solvers." *Arithmetic Teacher* (November, 1990), pp. 35–38.

Gere, Anne Ruggles, editor. *Roots in the Sawdust: Writing to Learn Across the Disciplines.* Urbana, IL: National Council of Teachers of English (1985), 1–8.

Gordon, Christine J., and Macinnis, Dorothy. "Using journals as a window on students' thinking in mathematics." *Language Arts* (January, 1993), pp. 37–43.

Graves, Donald H. *Writing: Teachers and Children at Work.* Portsmouth, NH: Heinemann (1983).

Helton, Sonia "I THIK THE CITANRE WILL HODER LASE: Journal keeping in mathematics class." *Teaching Children Mathematics* (February, 1995), pp. 336–340.

Hosmer, Patricia A. "Students can write their own problems." *Arithmetic Teacher* (December, 1986), pp. 10–11.

LeSieg, Theo. *Ten Apples Up On Top.* New York: Random House (1961).

Lewis, Barbara A., Long, Roberta, and Mackay, Martha. "Fostering communication in mathematics using children's literature." *Arithmetic Teacher* (April, 1993), pp. 470–473.

Matz, Karl A., and Leier, Cynthia. "Word problems and the language connection." *Arithmetic Teacher* (April, 1992), pp. 14–17.

McIntosh, Margaret. "No time for writing in your class?" *The Mathematics Teacher* (September, 1991),. pp. 423–433.

McGehe, Carol A. "Mathematics the write way. " *Instructor* (April, 1991), pp. 36–38.

Silverman, Fredrick L., Winograd, Ken, and Strohauer, Donna. Student-generated story problems." *Arithmetic Teacher* (April, 1992), pp. 6–12.

Stix, Andi. "PIC-JOUR MATH: Pictorial journal writing in mathematics." *Arithmetic Teacher* (January, 1994), pp. 264–269.

Vacca, R.T., and Vacca, J. *Content Area Reading.* Glenview, IL: Scott, Foresman (1989).

Wilde, Sandra. "Learning to WRITE about mathematics." *Arithmetic Teacher* (February, 1991), pp. 38–43.

Winograd, Ken. "Selected writing behaviors of fifth graders as they compose original mathematics story problems." *Research in the Teaching of English* (1993), Vol. 27, pp. 369–394.

Wood, Karen D. "Fostering collaborative reading and writing experiences in mathematics." *Journal of Reading* (October, 1992), pp. 96–103.

3

Creating an Appropriate Learning Environment for Writing in Mathematics

The classroom environment is foundational to what students learn and the teacher is responsible for creating an environment where everyone can do mathematics, engage in mathematical thinking, and solve mathematical problems.

TIME

A learning environment that fosters the development of each student's mathematical power is one where the teacher provides and structures **time** necessary for the exploration of mathematical ideas and problems by students. Writing in mathematics takes time—time for students to think, reflect, draw on their own personal experiences, manipulate materials, discuss concepts and understandings, and experience and explore mathematical concepts. Time must be allowed for students to explore, interact, and write.

Nancy Anderson, teacher of a K–1 multiage class, introduced her students to tangrams by reading *Grandfather Tang's Story*, by Ann Tompert, and to writing in journals in mathematics. After Nancy read the story to her students, she had them work in groups of four to discuss ideas from the story and describe different shapes that could be created using the seven tangram pieces. Then students explored designs and shapes using plastic tangram puzzles. At first students duplicated pictures from the book and from other patterns on tangram cards. Next, students created original pictures with tangrams and then orally composed stories about tangrams. On the following day, students worked in pairs. While one student created a design with tangram pieces, the partner copied the pattern. Nancy then read *The Napping*

House by Audrey Wood and Don Wood. After discussing the story, students used tangrams to make "houses."

All of the experiences with books, tangrams, and shapes invited communication and creative expressions in mathematics. Students had engaged in several activities that helped build mathematical understandings. The explorations with materials and discussions helped clarify students' reasoning. Students had explored mathematical concepts, communicated their thinking, and listened to ideas and thoughts of other students. Armed with these experiences, Nancy's K–1 students were ready to engage in journal writing. The prewriting experiences took time, a characteristic of an appropriate learning environment that fosters writing in mathematics.

Valerie Jones read *Anno's Mysterious Multiplying Jar*, by Masaichiro Anno and Mitsumasa Anno, to her fifth grade class. She and her students discussed the meaning of factorials and how they work. Valerie gave directions to students to write a parallel story. A parallel story is a story that follows a story line in the story read (in this case, *Anno's Mysterious Multiplying Jar*), but the parallel story varies in the characters, objects, and numbers included (Lewis, Long, and Mackay, 1993). Figure 3.1 shows one student's parallel story entitled, *Magical Mouse Hole*. The student spent a great deal of time on her story. She had to decide on an idea for her story. Then she had to make decisions about objects and numbers that would be appropriate for her story about the mouse hole. Next, the student had to compose and illustrate her story and calculate totals for the factorials for each page of her book. This entire process of making decisions, writing, illustrating, and publishing her story took several days. The student was given time by her teacher, time that is necessary to explore ideas for parallel stories. Valerie gave her students meaningful experiences where they did not just depend on textbooks but used an integrative approach by applying mathematical concepts to a children's story (Cangelosi, 1988), a strategy that takes time.

(Text continues on page 53.)

FIGURE 3.1. PARALLEL STORY ON FACTORIALS

There was one street....
(1! 1x1=1)

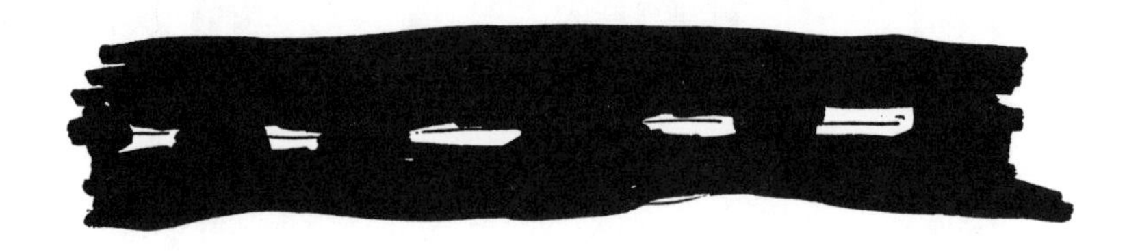

On that street there was two houses....
(2! 2x1=2)

In each house there were three rooms....
(3! 3x2x1=6)

In each room there were four mouse holes....
(4! 4x3x2x1=24)

In each mouse hole
there were 5 mice.....
(5! 5x4x3x2x1=120)

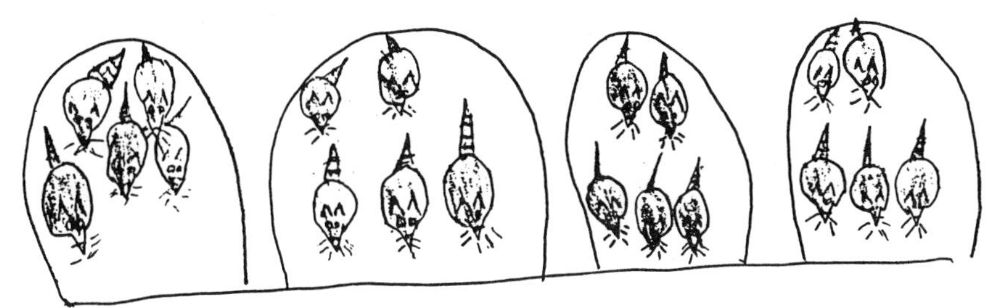

In each mouse there were
six peices of cheese......
(6! 6x5x4x3x2x1=720)

In each hole of the cheese there were 8 fur balls.....
(8! 8x7x6x5x4x3x2x1=40,320)

FOCUS OF THE LEARNING ENVIRONMENT

The learning environment for writing in mathematics is less teacher-directed and more focused on the students. Students are allowed to select their own writing topics, their own modes of writing and their own audiences for their writing. The environment is nonthreatening and anxiety-free, one where it is acceptable for students to make mistakes and to ask for help. The focus of the environment should be on the development of thinking skills and problem-solving, and one where discussion of the problem-solving process and strategies is highly valued. (Winograd and Higgins, 1994/1995; Sosenke, 1994/1995; NCTM, 1991.)

In this environment, writing is valued, encouraged, and supported. Questions are encouraged and discussion is an important part of both the problem-solving and writing processes. Students are encouraged to construct their own questions and responses. The environment is one in which students perceive they own the mathematics they have learned (Miller, 1993).

Al Boucher remarked that his first grade students were amazed at how he sat back somewhat and allowed them to talk to classmates to figure out solutions to real problems and to complete their writing and illustrations. Al stated that he accepted answers and solutions to problems as long as students were able to support their answers. Many times he noticed that students were able to correct their own errors and many offered to help their classmates. The nonthreatening, anxiety-free environment helps writing in mathematics take on new meaning. Al's first grade students were eager to draw pictures, write or dictate sentences about their mathematics problems, and were able to construct, record, and solve problems through their writing.

MATERIALS, ACTIVITIES, AND TASKS

Materials, activities, and tasks for classrooms that foster writing in mathematics must focus on problem-solving. For this focus, a wide variety of materials should be available for exploration and use by students.

Manipulatives of all kinds, including both commercial materials such as geoboards, Unifix cubes, tangram puzzles, attribute pieces, and counters, and materials from the environment, such as buttons, toy models, pencils, keys, and yarn, should be readily available for exploration by students. Students can be asked to describe in writing a manipulative, what they have discovered about the manipulative, and what they did with the manipulative. The frequent use of manipulative experiences to build a background for writing. (Mumme and Shepherd, 1990; Capps and Pickreign, 1993.)

A large variety of trade books, such as counting books, concept books, fiction and nonfiction books, as well as textbooks, should be available for exploration by students to generate ideas for writing in mathematics. The use of literature in a mathematics class creates opportunities for sharing, discussing, and writing (Kliman and Kleiman, 1992). (Appendix A is a bibliography of recommended books that encourage writing.)

Commercial, teacher-made, and student-made games are other sources of ideas for writing in mathematics. Especially appropriate for triggering ideas for writing in mathematics are games such as checkers, chess, and Tic Tac Toe. After students play games they can be encouraged to write step-by-step directions for playing a game to share with a friend. Students might also describe, in writing, strategies they use to win games.

Newspapers, pictures from magazines, and sale flyers contain information and ideas that students can discuss. They can compose and write their own mathematical stories and problems. Kay Stone selected a picture from a magazine that showed eight children standing in a row, smiling at the camera. Kay asked her third grade students to compose a story problem that contained mathematical information they found when they examined the picture. She discussed the picture with students and asked questions about the picture. She then asked students to write a mathematical story about the picture. One student composed a story about eight children at a party. The student wrote, "Eight children were at a party lined-up to have their picture taken. A bee buzzed by and chased three children off. Now there are only five children left for the picture. 8 – 3 = 5" (Fig. 3.2).

FIGURE 3.2. USING PHOTOS FROM MAGAZINES OR NEWSPAPERS

There are eight people on
this poster. If a bee came
and skerd some away.
There is five left.
How meny got skerd
away? 8-5=3

In another student's story problem about the same picture, each child in the picture had been given a name. The student asked, "Which child is first in line? Second in line?" etc. Kay was really pleased and somewhat surprised at the creativity of her students who wrote different mathematical stories about the same pictures. Different concepts were expressed in the stories.

Newspapers and magazines contain tables, charts, and graphs that elicit ideas for writing. Sale flyers that accompany newspapers or arrive in the mail contain excellent information that can serve as sources of activities and tasks that are prerequisites for writing in mathematics. Students can examine the tables, charts, and graphs, and sale flyers and compare the content or create original mathematics problems from information in the materials. Students can then write their own interpretations of the tables, charts, and graphs.

Menus from restaurants, train schedules, and supermarket ads also contain interesting, real-world information for students to use to build backgrounds, and enhance ideas and concepts necessary for writing. The use of a variety of materials and tasks that are adapted to the interest of students and tasks that encourage a variety of interpretations and strategies, are necessary for creating an environment for writing in mathematics. The tasks and activities using the materials described present a novel way and a variety of ideas for students to engage in investigations, to prepare and present projects, to create problems that reflect their everyday experiences, and encourage an integrated approach to learning.

QUESTIONING

The teacher's use of effective questioning techniques is important in an environment that encourages writing. Questioning is one of the most powerful tools to encourage students to interact with mathematical content. It is important that teachers ask questions that challenge students to make hypotheses and help students form categories. How teachers phrase questions is important and determines the kinds of responses given by students. Teachers need to use questions that encourage students to look for inferences, support ideas, and discover concepts and generalizations.

Asking students to write and reflect on their mathematical understandings and thoughts stimulates their thinking abilities. Responding to open-ended questions helps students use a higher level thinking.

Laurie King, a teacher of fourth grade students, was concerned that the type of questions she created for her students had a direct effect on how her students were developing their reasoning skills. She examined a geometry activity sheet she was using with her students. Laurie admitted that even though she asked many questions, the majority of her questions fell into the "factual" category. She feels she will gain only limited information about her students' reasoning and mathematical abilities. Laurie's school uses the *Everyday Math* program from the University of Chicago. She feels this program will help her in developing higher level questions for her students. Laurie evaluated her questions, and then she made a new set.

Laurie's Original Questions:

1. Which of these shapes are polygons? (factual)
2. Which of these shapes are parallelograms? (factual)
3. How are shapes "L" and "U" alike? Different? (open/reasoning)
4. Which of the shapes are triangles? (factual)
5. Measure the angles in Shape "A." (factual)
6. Name two line segments in Shape "S." (factual)
7. How are Shape "G" and "I" different? (open)
8. Which shape is a rhombus? (factual)
9. Why is Shape "C" a concave polygon? (open)
10. What is the shortest line segment of Shape "S"? (factual)

Laurie's New Questions:

1. Approximately how many times would shape "B" fit into "A"?

2. Which of these figures could be divided into 4 equal parts?
3. Which of these shapes cannot be divided equally?
4. Why is shape "O" not a circle?
5. Does shape "T" have any obtuse angles?
6. In what ways are shape "M," "O," and "E" different?
7. How are shapes "U" and "T" alike?
8. How many "L" shapes would it take to make a rectangle with the same width as "U"?
9. Which shape doesn't belong: A,B,H,N,R,S,K,F ?
10. How would you describe shape "C"?

Laurie concluded that her new questions would elicit higher-level responses from students and would enhance their reasoning abilities. She plans for her students to respond both orally and in writing to new questions in geometry and other strands in the mathematics program.

ROLE OF THE TEACHER IN THE ENVIRONMENT

The teacher plays an important role in creating an environment that encourages writing in mathematics. It is important that the teacher validate and support the thinking and ideas of students. The teacher is a facilitator, a guide, the one who creates an environment that activates discovery and fosters learning. The teacher provides many opportunities for students to communicate their thinking about mathematics through reading, writing, speaking, and listening. The teacher encourages students to develop their own problems and allows them to develop their own methods of investigation. Discussion and questions are encouraged as the teacher validates and supports students' ideas. (Olson, 1984; Mumme and Shepherd, 1990.)

The teacher who uses writing in mathematics has a good understanding of her students, their interpretations, experiences, cognitive styles, and developmental levels. She shows that she values mathematics and writing in mathematics and is not judgmental or evaluative of responses (Gordon and Macinnis, 1993).

The teacher who provides an appropriate environment for writing

- Models appropriate vocabulary with students;
- Encourages students to conceptualize situations in alternative ways to encourage them to develop their own problems (Miller, 1993);
- Uses natural prompts and presents both overt and covert cues to students;
- Shows she values writing by modeling writing skills with students and making sincere responses to students' writing; and
- Uses physical space and materials to facilitate students' learning of mathematics (NCTM, 1989, p. 57).

The environment or tone of the classroom that is appropriate for writing in mathematics is set by the teacher and what she does, not what she says. Students and teachers need time: students need time to become discussants, writers, and solvers of problems; teachers need time to become accepting of the ideas and contributions of students' writing problems and stories, and time to become accustomed to this approach in teaching and learning mathematics.

ROLE OF THE STUDENT

In an environment that supports writing in mathematics, students are actively engaged in their own learning. They are encouraged to use their own initiative in figuring out problem solutions for themselves. Students contribute their own ideas and generate questions about mathematical problems. They ask questions of others and of the teacher. They offer suggestions of ideas and critiques of the ideas of others. Writing and oral responsibilities of students in an environment that promotes writing as a means of learning mathematics includes:

- Cooperating with other students;
- Generating ideas;
- Interacting with other students;

- Interacting with the teacher;
- Asking questions;
- Responding to questions and ideas from other students and the teacher;
- Conducting mathematical investigations;
- Thinking about alternative solutions;
- Discovering answers to questions through investigations;
- Reflecting on their own mathematical understandings and feelings about mathematics;
- Challenging what is presented; and
- Becoming an audience for other students' writings.

REFERENCES

Cangelosi, James S. "Language activities that promote awareness of mathematics." *Arithmetic Teacher* (December, 1988), pp. 6–9.

Capps, Lelon R., and Pickreign, Jamar. "Language connections in mathematics: a critical part of mathematics instruction." *Arithmetic Teacher* (September, 1992), pp. 8–12.

Gordon, Christine J., and Macinnis, Dorothy. "Using journals as a window on students' thinking in mathematics." *Language Arts* (January, 1993), pp. 37–43.

Kliman, Marlene, and Kleiman, Glenn M. "Life among the giants: Writing, mathematics, and exploring Gulliver's world." *Language Arts* (February, 1992), pp. 128–136.

Lewis, Barbara A., Long, Roberta, and Mackay, Martha. "Fostering communication in mathematics using children's literature. *Arithmetic Teacher* (April, 1993), pp. 470–473.

Miller, L. Diane. "Making the connection with language." *Arithmetic Teacher* (February, 1993), pp. 311–316.

Mumme, Judith, and Shepherd, Nancy. "Communication in mathematics." *Arithmetic Teacher* (September, 1990), pp. 18–22.

National Council of Teachers of Mathematics. *Curriculum and Evaluation Standards for School Mathematics.* Reston, VA: The National Council of Teachers of Mathematics (1989).

National Council of Teachers of Mathematics. *Professional Standards for Teaching Mathematics.* Reston, VA: The National Council of Teachers of Mathematics (1991).

Olson, Carol Booth."Fostering critical thinking skills through writing." *Educational Leadership* (November, 1984), pp. 28–39.

Sosenke, Fanny. "Students as textbook authors." *Mathematics Teaching in the Middle School* (September/October, 1994), pp. 108–111.

Winograd, Ken, and Higgins, Karen M. "Writing, reading, and talking mathematics: One interdisciplinary possibility." *The Reading Teacher* (December, 1994/January, 1995), pp. 310–318.

CHILDRENS BOOKS

Anno, Masaichiro, and Anno, Mitsumasa. Anno's Mysterious Multiplying Jar. New York: Philomel Books (1983).

Tompert, Ann. Illustrated by Robert Andrew Parker. *Grandfather Tang's Story.* New York: Crown Publishers (1991).

Wood, Audrey, and Wood, Don, ill. *The Napping House.* San Diego, CA: Harcourt (1984).

4

Writing to Create and Solve Problems

As early as 1980, the National Council of Teachers of Mathematics stressed that the primary focus of school mathematics is to develop students' abilities to solve problems occurring in the real world (NCTM, 1980). According to the K–4 and 5–8 Standards, the study of mathematics should emphasize problem-solving so the students can:

- Use problem-solving approaches to investigate and understand mathematical content;
- Formulate problems from everyday and mathematical situations, from within and outside mathematics;
- Develop and apply a variety of strategies to solve a variety of problems;
- Verify and interrupt results with respect to the original problems; and
- Acquire confidence in using mathematics meaningfully (NCTM, 1989, pp. 23 and 75).

Mathematics should be communicated in writing because writing in mathematics promotes mathematical learning. Writing can be used as a strategy for making the process or activity of problem-solving meaningful to students. A problem-solving strategy that appears to be highly successful with mathematical word problems is to have students write their own problems (Ford, 1990; Fennell and Ammons, 1985). Until students formulate their own word problems, they will not fully understand the reading contained in mathematics. When students pose their own problems, their understanding of problem-solving is enhanced, they learn to generalize, and mathematics becomes more meaningful (Graves, 1978; Brown and Walters, 1983; Wirtz and Kahn, 1982).

PROBLEM-SOLVING AND PROBLEM-POSING

Problem-solving is a process that should permeate throughout the mathematics program and should be the context in which concepts and skills are learned (NCTM 1989, p. 23). Writing is a problem-solving process in which problem finding is an integral part. Before students engage in writing to solve and pose word problems in mathematics, they need many and varied experiences. Students must have opportunities to interact with their peers and teachers in generating ideas. They need time to reflect on previous experiences, to find and define problems to be solved, and to find opportunities to participate in activities that build a background for problem-posing and problem-solving. Students should be exposed to many models of the kinds of problems we expect them to write. (Winograd, 1993; Cambourne, 1988.)

Angela White believes the most worthwhile activity she does with her second grade students is when they write their own story problems. She indicates that she sees growth in students' problem-solving skills in mathematics and in their story problem writing. She uses a picture for a group discussion and then students work in groups to write a word problem about the picture.

USING THE WRITING PROCESS WITH MENUS

Several teachers have used restaurant menus with their students to help them create a variety of realistic, everyday word problems. Teachers (third, fourth, and fifth grades) made calculators available to students so the students could concentrate on creating and writing mathematics problems. Teachers gathered menus, distributed them to their students, discussed the makeup of the menus, the prices, taxes, tips, and making choices. Then students made suggestions about what foods they liked to eat appearing on the menus. Next students composed, in writing, their own story problems based on their food choices and the prices on the menus.

Figure 4.1 shows one student's problem story. He did not indicate which menu he used, but did write an appropriate story with a mathematical content. His question, "If I had $20.00, can I buy 2 more sodas for $1.00 (dollar)?" He did not list prices in his story problem.

FIGURE 4.1. CREATING PROBLEMS FROM RESTAURANT MENUS

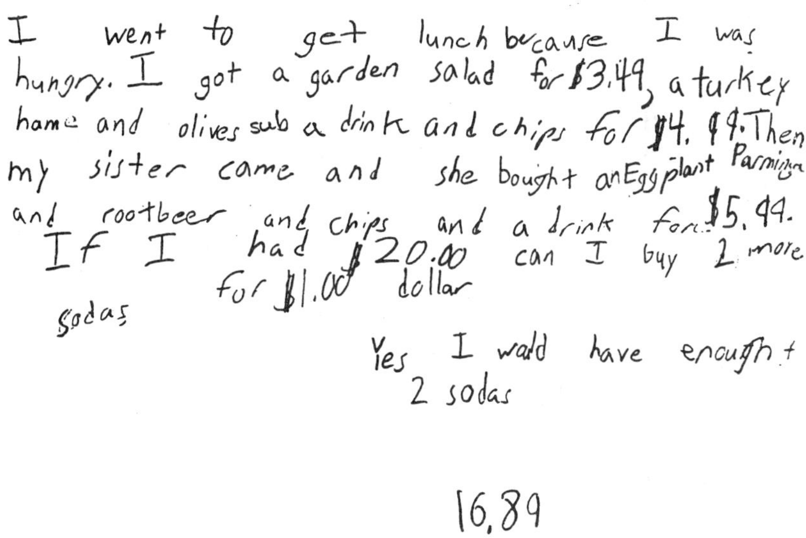

I went to get lunch because I was hungry. I got a garden salad for $3.49, a turkey hame and olives sub a drink and chips for $4.99. Then my sister came and she bought an Eggplant Parmigna and rootbeer and chips and a drink for $5.44.
If I had $20.00 can I buy 2 more for $1.00 dollar sodas

Yes I wald have enougth t 2 sodas

16.89

An appropriate step for students using the writing process to create word problems, is for students to exchange problems with a partner for editing. Each student reads their partner's problem and addresses these questions:

- Does the story make sense?
- Is there a problem to be solved?
- Is the necessary information for solving the problem included?
- Is an appropriate question asked?
- Is the content of the story problem reasonable?
- Does the story problem need corrections of spelling or punctuation?

If the student has solved the problem posed, ask:

- Is the solution correct?
- What suggestions can you offer to your partner for creating story problems or finding the solution to his or other problems?

Next students work on their own story problems to make corrections, to edit them, and to do any rewriting that is necessary. Finally, students publish their problems by writing them on cards to be placed in a Mathematical Problems Box or record and illustrate them on chart paper or a poster. These can be placed on the wall or on a bulletin board where other students can read and solve them.

Wendy Wicht "gave" her students $25.00 to buy dinner for themselves and three of their friends. When students examined menus and discussed prices, they realized that it was not possible for four people to eat steak on only $25.00. Some students suggested that each of the four students in their group should spend the same amount, but when students made the selections, most chose to spend more for their own food than for their friends.

When Wendy's students completed their food selections, and before they calculated the total to see if they could purchase the selections, each student made a chart to show selections, estimated cost, and actual cost. One student's chart is shown on the next page.

Students wrote three word problems related to their selected dinners. The students whose selections are shown in the chart wrote these questions:

- After I pay my total bill, will I have enough to order apple cobbler?
- What is the price difference between the steak and the pork sandwich?
- Could I buy another steak?

In this experience, students used oral communication, reasoning and estimation, as well as writing to solve problems.

Student	Order	Estimated Cost	Actual Cost
Student 1	Brunswick Stew, Iced Tea	$3.00	$1.85 .75
Student 2	Chicken Tenders, Salad, Water	5.00	5.05
Student 3	Tossed Green Salad, Water	1.00	1.25
Student 4	Steak, Milk	12.00	10.95 .80
		21.00	20.65
	Tip		3.10
	Total		$28.75

(Students calculated a 15% tip for each order but did not include taxes.)

Karen Miller, another third grade teacher, used takeout menus with her students after she read an article by Ford (1990). She visited several restaurants, collecting takeout menus. She reviewed steps in the writing process with her students. She then showed students the examples Ford used in her article and students discussed the writing process and problem-solving process.

Karen then had students individually write first drafts or "sloppy copies" of story problems on unlined paper. Students then worked with a partner and read each other's problems, and together worked on "improving" them. Karen also read the problems and helped students with their final editing. She did as Ford recommended. Students wrote their finished problems on colored file cards and placed them in a special box which was placed in their math center.

Karen says the problems created in this "first time" activity were not very challenging or stimulating, and students used calculators. She was pleased enough with the word problems to try the activity again with her students. She invited students to add more story problems to their box when they chose. Karen believes the quality and complexity of problems will increase as students become accustomed to using the writing process.

Figure 4.2 shows a fourth grade student's final story problem, the first she had created. Her teacher did not supply menus for the class, but allowed her students to decide on the price of items. The student worked hard on making her food problem "neat." The problem was not edited. Since this was the student's first attempt at creating her own problems, she was pleased.

A fifth grade student wrote the story and accompanying problem shown in Figure 4.3. It is evident from examining her problem that she was not comfortable writing on unlined paper and that she spent almost as much time drawing lines and illustrating her problem as she did in creating the story.

The story problem was not edited and did not contain information needed to answer the question. The author of the problem used her illustrations as a way of presenting necessary information. From her calculations it appears she purchased a pizza for $6.55 from Pizza Hut, "liman" from Chick-fil-A for $2.00 and a drip "cron" from Dairy Queen for $3.95. She posed the question, "The tata of money is?"

This story problem is a beginning for the student and her teacher. I recommended to the teacher that she place less emphasis on the illustrating of story problems and more on working with students in learning to create and write good situations that contain all the information needed for posing problems. I also suggested that she read the Ford article and related articles in which the writing process is used, and to try these ideas with her students.

PICTURES AS SOURCES OF PROBLEMS

Jill Fowler had her students "try their hand" at writing word problems. Her second grade students had previously had difficulties in solving word problems. She stated that even when students gave correct answers to word problems, she wondered if they really understood the problems and why they solved them

FIGURE 4.2. FOOD PROBLEM

Michael and Natashia have 4 dollars. Michael and Natashia went to the store and they bought 2 dirnks and 2 Bags of chips and 2 candy bars. How moch did it coughts. Addto find the answer.

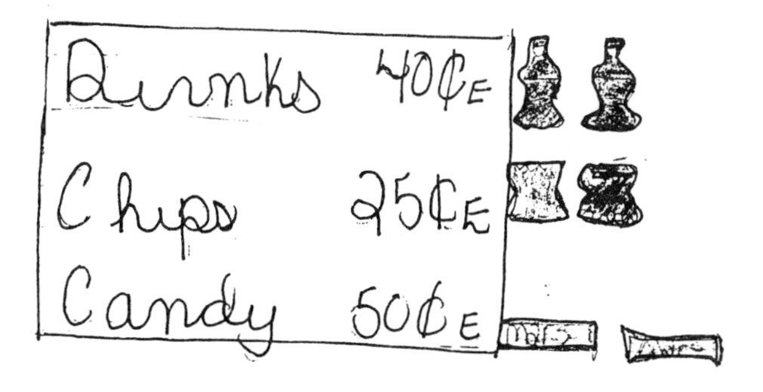

FIGURE 4.3. BEGINNING STORY PROBLEM FOR STUDENT AND TEACHER

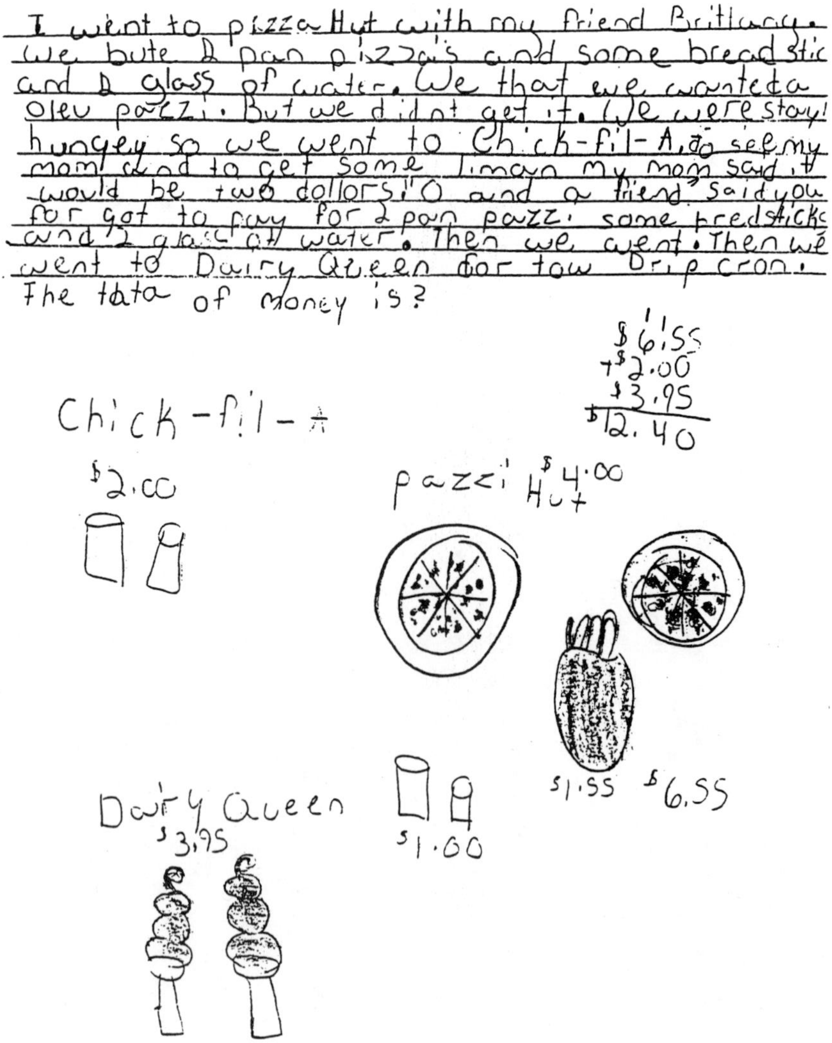

I went to pizza Hut with my friend Brittany. We bute 2 pan pizza's and some breadstic and 2 glass of water. We that we wanted a olev pazzi. But we didnt get it. We were stay! hungey so we went to Chick-fil-A, to see my mom and to get some limaon my mom said it would be two dollors!O and a friend said you for got to pay for 2 pan pazzi some bredsticks and 2 glass of water. Then we went. Then we went to Dairy Queen for tow Drip cron. The tata of money is?

$6.55
+$2.00
$3.95
$12.40

Chick-fil-A
$2.00

pazzi Hut $4.00

$1.55 $6.55

Dairy Queen
$3.95

$1.00

in the manner they did. With these thoughts in mind, Jill put her students into groups to write and solve word problems.

I suggested to Jill and to other teachers that they find and use pictures from magazines, such as *Good Housekeeping, Family Circle, Woman's Day,* or similar magazines that had pictures with which students would be familiar. Another suggestion offered was to use a poster as a way of providing ideas for students to create and solve real problems. Jill selected and cut 20 pictures from magazines.

Jill then put her students into groups of three or four and had each group select a picture. Once students were in groups, they described items in the selected picture and began to generate information from the pictures that contained numbers or other mathematics content. Many groups changed pictures several times before they were ready to begin writing story problems. It was important that the groups were comfortable with the pictures and felt there were enough mathematics ideas that could be generated from them.

Jill asked that each group write a story and a numerical question to accompany the story. She overheard several students questioning members of their own group about whether or not their problem made sense. When the groups presented their stories and problems to the rest of the class, some stories and problems did not include enough information for an answer or solution to be found for the questions posed. These groups had to "regroup" and reedit and rewrite their story problems. Eventually, each group created a story and posed a problem from the picture they had selected.

When all stories and problems were corrected, Jill's students glued the pictures to the front of a sheet of manila paper and wrote the story problem below the picture. Students wrote an equation that could be used to solve the problem on the back of the manila paper.

Jill's students then drew pictures and wrote explanations of their problems, describing in writing how to solve the problems. She felt that this writing activity enriched her students' understanding of word problems and they were better able to understand questions that are asked in word problems. As a regular part of her mathematics program, Jill indicates she puts a word problem on the chalkboard each morning. Her students seem to look forward to the problems and are eager to attempt to solve them before school each morning. Jill plans to start a collection of original word problems written by students and make them into a class book to be used throughout the year.

Meredith McTyre also used pictures cut from magazines as a basis for story problems. Her third grade students enjoyed the activity and some students who were usually not successful in solving word problems from a textbook were successfully able to compose and solve original mathematics word problems. Meredith indicated that as she observed her students writing she realized difficulties students were having and she could readily assess what they understood about the problems and problem-solving in general. She felt that writing problems from pictures was a worthwhile activity and she will continue to use them in problem-solving activities.

Figures 4.4 and 4.5 are examples of story problems created by individual second and third grade students. The problem posed in Figure 4.5, "How many in all?," is a typical textbook-type question, as is the question posed in Figure 4.4, "How many were left?" Both questions contain what appears to be clue words, "in all" and "left." Students who have many experiences with writing more realistic problems eventually will expand their understandings and ask questions that do not contain clue words.

The question asked for the picture and story in Figure 4.6 is a more realistic question, something teachers should help their students understand. "What is the other adden(d)?" is the question in Figure 4.7. It appears the student who composed this story and question, is familiar with the concept of addends.

The student's question in Figure 4.8, "How much flowers did she water?," indicates that she understands the concept of adding and is beginning to develop terminology appropriate for the problem situation.

As students continue to compose original problems, they will have a better understanding of appropriate questions for their story problems.

(Text continues on page 79.)

FIGURE 4.4. STORY PROBLEM WITH MAGAZINE PHOTO (I)

FIGURE 4.5. STORY PROBLEM WITH MAGAZINE PHOTO (II)

Thir was two football teams. One team had three player the other team had three players. How many players in all.

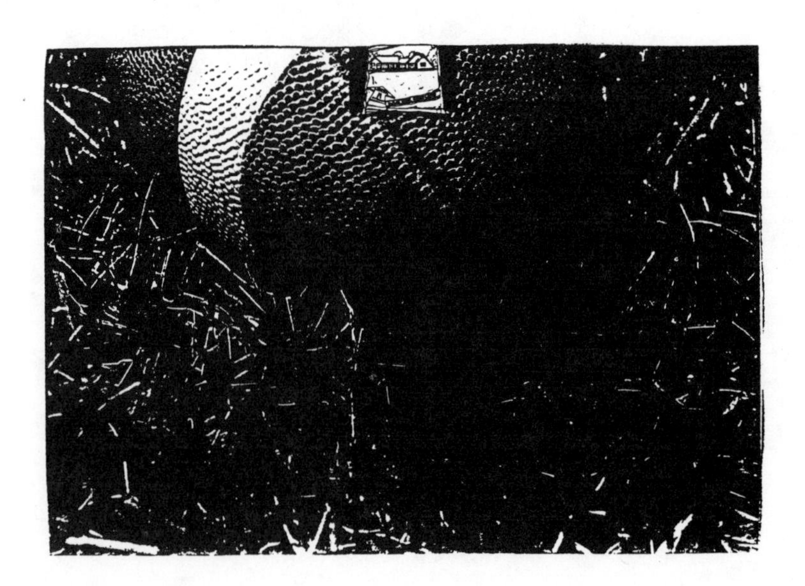

FIGURE 4.6. USING A MORE REALISTIC QUESTION

candy weighed 200 Pounds. She loss 63 pounds.
how much does candy weigh: ? ________

FIGURE 4.7. BECOMING FAMILIAR WITH MATHEMATICAL CONCEPTS

FIGURE 4.8. BEGINNING TO LEARN MATHEMATICAL TERMINOLOGY

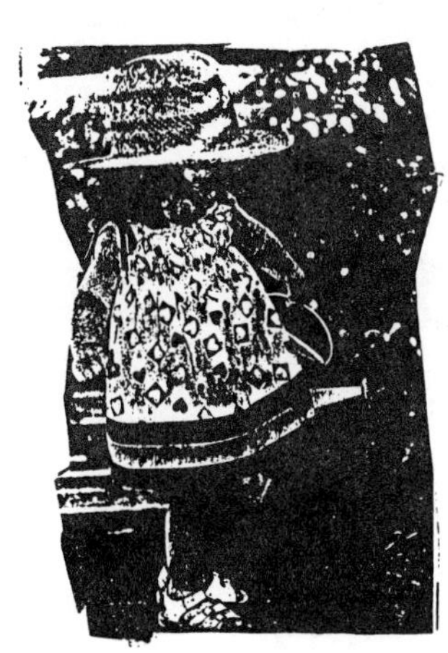

Amanda waterd 27 flowers the first day. The nexet day she watered 15. How much flowers did she water.

PROBLEM-SOLVING AFTER LISTENING TO A CHILDREN'S BOOK

Nancy Anderson read *Six Brave Explorers*, by Kees Moerbeek, to her first grade students. They discussed whether the story had anything to do with mathematics. Students acknowledged that the story contained number words, sets to count, and each page in the book was really a subtraction problem. Nancy placed her students in small groups and gave them bear counters. Each group used the counters to illustrate the story as they retold the story in their group. As one student retold the story, the other members of the group decided whether the incident illustrated with counters was an addition or a subtraction problem.

This activity, following the reading of a story, enhanced the oral communication skills in mathematics and helped increase students' understanding of problem posing and solving. The activity also gave students experiences necessary for writing word problems.

The following day, Nancy paired her first grade students with a second grade "buddy" for the purpose of writing story problems. Nancy's class was studying a unit on the rain forest and she had read *Rain Forest,* by Helen Cowcher, to her students. After hearing the story, students examined pictures in the book and, as a group, students orally composed word problems. Nancy recorded the word problems on the chalkboard where they served as models for individual students. Students decided they wanted to make their own word problems and illustrations based on the story of the rain forest. They wanted to write the problems and have students in another class read and solve their problems.

Student buddies wrote drafts of their stories and exchanged their story problems with other students in their class to make sure the problem they had written made sense and that the reader understood what was being asked. After the exchange, students revised their stories when necessary. When the stories were completed, students decided to send their book of stories to other first and second grade classes for them to solve.

There are many kinds of writing activities that can be used to help students become problem-solvers. In addition to those described, students can write explanations of how they solved problems. These explanations are often recorded in logs. Writing about what they have learned in mathematics and writing explanations of how problems are solved are other ways that students can become problem-solvers. Logs and writing explanations are described and illustrated in later chapters.

REFERENCES

Brown, Stephen I., and Walter, Marion, J. *The Art of Problem Posing.* Philadelphia: Franklin Institute Press (1983).

Bush, William S., and Fiala, Ann. "Problem stories: A new twist on problem posing." *Arithmetic Teacher* (December, 1986), pp. 6–9.

Cambourne, Brian. *The Whole Story: Natural Learning and the Acquisition of Literacy in the Classroom.* Auckland, NZ: Ashton Scholastic (1988).

Fennell, Frances, and Ammon, Richard. "Writing techniques for problem solvers." *Arithmetic Teacher* (September, 1985), pp. 24–25.

Ford, Margaret I. "The writing process: A strategy for problem solvers." *Arithmetic Teacher* (November, 1990), pp. 35–38.

Graves, Donald. *Balance the Basics: Let Them Write.* New York: Ford Foundation (1978).

Helton, Sonia H. "I THIK THE CITANRE WILL HODER LASE: Journal keeping in mathematics class." *Teaching Children Mathematics* (February, 1995), pp. 336–340.

Kliman, Marlene, and Richards, Judith. "Writing, sharing, and discussing mathematics stories." *Arithmetic Teacher* (November, 1992), pp. 138–141.

National Council of Teachers of Mathematics. *An Agenda for Action: Recommendations for School Mathematics of the 1980's.* Reston, VA: National Council of Teachers of Mathematics (1980) .

National Council of Teachers of Mathematics. *Curriculum and Evaluation Standards for School Mathematics.* Reston, VA: National Council of Teachers of Mathematics (1989).

Rickles, Fred. "Student-generated story problems." *Arithmetic Teacher* (April, 1982), pp. 6–12.

Winograd, Ken. "Selected behaviors of fifth graders as they compose original mathematics story problems." *Research in the Teaching of English* (December, 1993), pp. 369–394.

Wirtz, Robert, and Kahn, Emily. "Another look at applications in elementary school mathematics." *Arithmetic Teacher* (September, 1982), pp. 21–25.

CHILDREN'S BOOKS

Crowcher, Helen. *Rain Forest.* New York: Farrar, Strauss, and Giroux (1988).

Moerbeek, Kees, and Carla, Dijs. *Six Brave Explorers.* Los Angeles, CA: Price Stern Sloan Pub. (1989).

5

Writing to Communicate Mathematics

By communicating mathematics, students are able to remember, understand, and use the information, and this leads to finding more information. They begin to construct links between intuitive notions and symbolism and make connections between "physical, pictorial, graphic, symbolic, verbal and mental representations of mathematical ideas" (NCTM, 1989, p. 26). Writing should be an integral part of the mathematics curriculum and is particularly useful in mathematics. When students write their thoughts they are able to clarify their ideas (Perkins, 1992).

The American Association for the Advancement of Science supports using written communication as an effective teaching strategy for mathematics and describes effective teaching of mathematics as teaching that emphasizes the development of students' abilities to communicate. Written communication is a tool that allows students to make connections in mathematics, reflect on their understandings of mathematics. Writing helps personalize mathematics. (AAAS, 1988; Buschman, 1995.)

Mumme and Shepherd authored an outstanding article, "Communication in Mathematics," in the *Arithmetic Teacher* (September, 1990). In the article, they discussed five functions of communication in mathematics. Two of the functions are especially relevant to this chapter's content, writing to communicate: "[Writing] helps students enhance their understandings of mathematics and [writing] can empower students as learners."

In this chapter many projects done by teachers, as well as works of students, are presented. The projects reflect the teachers' philosophies and the student's works show different developmental levels in mathematics and various stages in writing.

JOURNAL WRITING

The most effective method of using writing to teach mathematics is through a journal. Students record their thoughts and ideas and the teacher responds to what the student has written in the journal. A journal is a written conversation or talk on paper. Through journals, teachers begin to understand better what is happening in the classroom by "listening" and observing in a more focused way. The use of prompted writings in mathematics where the teacher poses a question or supplies a beginning of a sentence are recommended as well as free-writing where students write openly, because this type of writing encourages the discovery of ideas. (Vacca and Vacca, 1986; Gordon and Macinnis, 1993.)

Students can begin writing in journals to record their understandings and illustrate concepts as early as kindergarten. Figure 5.1 shows a Kindergarten student's entry in her mathematics journal. She simply wrote one sentence—"I teck math is esy"—and illustrated her entry.

Other kindergarten journal entries are shown in Figure 5.2. For each of these entries, the teacher gave a prompt or beginning. The prompt for one was, "One day at school I." The student replied that she "found six crayons at my table." She illustrated her response which was given orally to her teacher.

Figure 5.2 also shows an illustration of the kindergartner's response to "So I." The student responded orally, "Got one more because I needed a black" and drew herself and six crayons on one side and one crayon on the other side. The response to "Now I" was "have seven crayons to draw a dog" so the student drew herself and the seven crayons.

Nancy Anderson believes that a math journal works well with her class. She wrote:

> My class of 22 first graders have begun a math journal to record their feelings about math. It will also serve as a log of their math understandings. It will give them the opportunity to develop written communication in math (standard 2) and will be another tool for me to assess their math skills and reasoning.

FIGURE 5.1. KINDERGARTEN MATHEMATICS JOURNAL

I TECK MATH IS ESY. Good!

FIGURE 5.2. RESPONDING TO PROMPTS

The children began by helping me to construct the journals and designed their covers. I let them use their problem-solving skills by brainstorming what topics are included in the math content area. The children then had the chance to use any illustrations that would demonstrate that this journal was math related. The children used pictures of shapes, numbers, calculators, clocks, patterns, and number sentences. The children concluded the lesson by writing their first entry about how they feel during math time. They were to include what things they liked or disliked about math.

This project will continue on a weekly basis, allowing for a minimum of two entries a week. These entries will include a combination of logs on new learnings and feelings about math.

Keeping a journal in mathematics is one way children can express and document their mathematical thoughts. Journals allow children to explore the understanding of a mathematical concept and to communicate those understandings through writing. (Helton, 1995, p. 336.)

The most important mode for students' writing about mathematics is open-ended writing following a lesson. Students write about what they did or what they learned during the lesson. (Wilde, 1991.)

Three entries from a first grade student's math journal is shown in Figure 5.3. Kindergarten and first grade students can begin to express their feelings and describe their understandings in a mathematics journal as is also shown in Figure 5.4. The first grade student illustrated his understanding of adding (1 pumpkin + 1 pumpkin = 2, and 2 apples + 2 apples = 4).

Reflective writing in journals is when students simply write to make sense of and give shape to what they have learned about how they learned (Sanford, 1988). It is believed by many that knowledge is constructed when one "thinks" and reflects on the thinking. When students reflect in journals, they again are making the learning and knowledge their own. It is also important to respond in writing to students' reflective writing in journals.

Dorothy Glass had her third grade students write in reflective mathematics journals. She indicated that when she read the students' journals she could see what students knew, and if their concepts were incorrectly formed. She stated that writing in journals enabled her students to reflect on the concepts they had learned and her students felt that writing in mathematics journals was refreshing.

Figure 5.5 (shown on page 93) shows three entries in a third grade student's journal. She responded to prompts or suggestions supplied by her teacher, Karen Miller. In the first entry, the student explains that she likes math; in the second she explains how math is used by her mom; and in the third entry she describes an activity that involves estimating and making "munch mix."

A fourth grade student responded in her journal to the topic of "What do you know about multiplication?" (Fig. 5.6, shown on page 94). It is evident from reading the student's response that she has a beginning concept of multiplication. She indicated that multiplication is hard and that she first started multiplying the previous year.

FIGURE 5.3. THREE ENTRIES FROM A FIRST GRADER'S MATH JOURNAL

I Learned a bout Greater Date 10-6-94
or Less then. It was fun and Very
hard I just got finish. I leared
how to do math to.

I leaned in math Date 10-10-94
today. How to pit things in
ordar it was Called ordinal
nubers. Up front. Of the room
my teacher pit thing in
ardor.

Today in math I Date 10-11-94
leared how to do a coulonder.
I did the days of the week.
It was not hard eather.

FIGURE 5.4. ILLUSTRATED ENTRY FROM A FIRST GRADER'S MATH JOURNAL

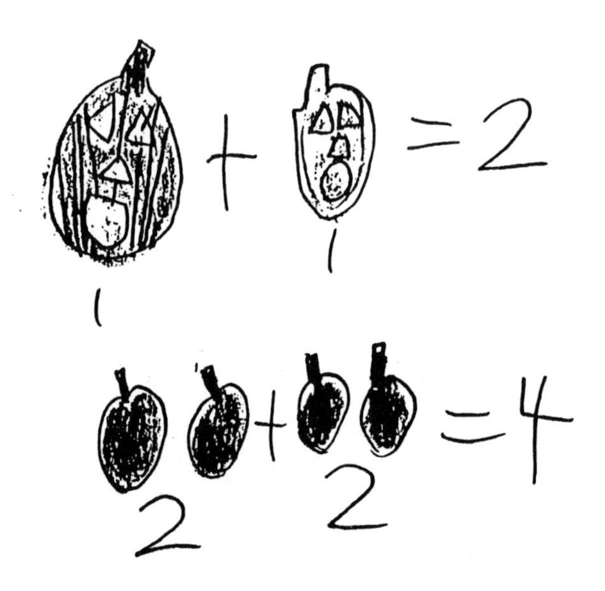

A fifth grade student, really responded to the prompt, "What I like about math." It seems that she likes most everything in math (Fig. 5.7, shown on page 95) she does, except "some of the long pages."

Tammy Cochran, a third grade teacher, began mathematics journals with her students. She stated, "I like writing in math. I was able to see if my students understood the math concepts by their responses in their journals. At first, my students were unsure how to respond in journals, but after working with them for a couple of weeks, they became more secure."

Jill Fowler reiterated those same feelings. She stated that her students enjoyed working in their math journals and she enjoyed the journals because they helped her see who really understood the concepts being addressed.

Kym Eisgruber had this to say about using journals in mathematics:

(Text continues on page 96.)

FIGURE 5.5. THREE ENTRIES FROM A THIRD GRADER'S MATH JOURNAL

There is not anything that I don't like about math because math makes your brain think better like when your in the store you need to know if somebody ask you to give them $15.39 you not what to give them because you will not know how to add.

10

My mom uses math when shes on the CCT bus. My dad uses math when he consentrates on reding books because he always counts his books to see how many books he has.

In yesterdays math we made munch mix. We had to estmate how many bags of munch mix we had. Another thing we did in math was we had to measure to make the mix.

FIGURE 5.6. RESPONSE TO "WHAT DO YOU KNOW ABOUT MULTIPLICATION?"

Multiplication

Multiplication is when you try to multply something, for example you have three pennies and two more sets of three thats nine so thats how you Multiply.

Multiplication is kind of hard to me but most of the time it is easy for me.

Most of the time Multiplication is fun because I love to write and draw.

I first started doing Multiplication in third grade! Not in eny other grade!

FIGURE 5.7. RESPONSE TO "WHAT I LIKE ABOUT MATH"

What I like about math

I like studying math problems like lenth, addition, subtraction and multiplication. I like the math book. I like the work sheets and the fun games. I like working together in groups and I like measureing things alot. I like math because it's at the end of the day also. I like writeing problems and I like homework I like the scales and I like when we use stuff that we eat like the weight or mass or Volume or matter. I like adding and subtracting. Some people like math like I do and some people don't like math and I ask them "Why don't you like math?" Somethings I don't like math like some of the long pages I do sometimes. I want to do the fun work sheet. I like doing telling time sometimes. last grade I got was 100 I think Math is easy so I always do my best I like writeing this page. I still like math. Well I like to use the metric and cusamary systems.

I decided to integrate journals and math not only because one of our school's goals is to improve in these areas, but because our students needed a bit more excitement when working with these subjects. Journals allow for flexibility in the classroom. They also help in the way that it is fun in school and you can send them home to keep the communication with the parents. The journals are not intended to replace our math series but to extend and supplement our existing curriculum. It does not require the purchase or use of large, expensive equipment to implement the use of journals. All that is needed are pencils and paper!

Incorporating math journals together has helped me to relax and let my children "experience" math and find their own answers. It helped me to take a step back and evaluate my role as the teacher. I realized that teaching is not only conducting a lesson in the classroom, but also acting as an observer. It is apparent that this role really helps the children's attitudes towards the subject. After all, nobody likes to get the wrong answer. So I had to ask myself, what is school all about?

COMMUNICATING THROUGH LETTERS

Students can write letters about what they have learned in mathematics or to express their feelings about mathematics. They may write a letter to tell a friend about an experience in mathematics. A student wrote a letter to another student (Fig. 5.8), in which he describes what he likes most about math and gives an example of a "hard" math problem.

Figure 5.9 is a letter written by a third grade student to a friend expressing her feelings about mathematics. She illustrated examples on a chalkboard using plus, minus, multiplication, greater than, and less than symbols. The examples she included show that she understands the symbols she wrote about.

FIGURE 5.8. LETTER TO A FRIEND

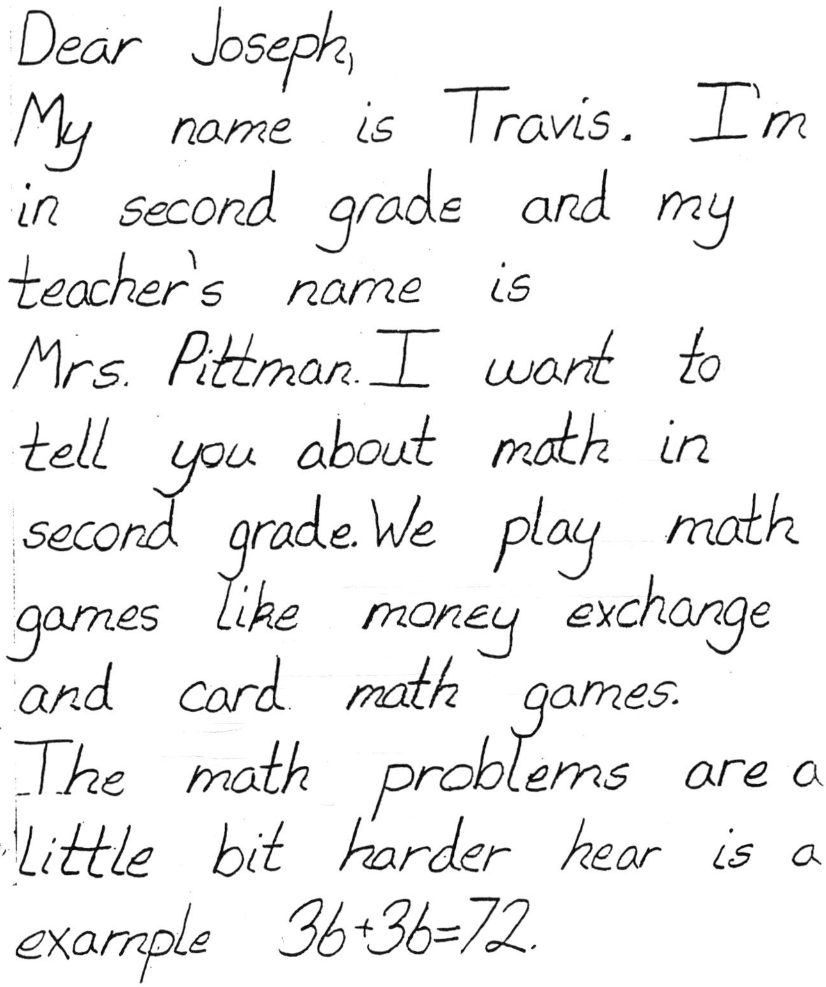

Dear Joseph,
My name is Travis. I'm in second grade and my teacher's name is Mrs. Pittman. I want to tell you about math in second grade. We play math games like money exchange and card math games. The math problems are a little bit harder hear is a example 36+36=72.

FIGURE 5.9. ILLUSTRATED LETTER

Dear Latosha, 11-21-94.
3-B.

When you get in third grade you will do math. It will be hard and it will be easy. You will do −, +, x and >, <. You might think it is hard but it is easy. I will draw −, +, x and π on the board.

Love,
Scheylla.

Valerie Jones' fifth grade students wrote a letter about themselves explaining how they were mathematicians. The student author of Figure 5.10 does not explain how she is a mathematician, but her description indicates that she thinks mathematics is important and that a mathematician works hard. Her response to the assignment and those from other students could be the basis for a lesson in which students explore the lives of mathematicians. Valerie might have her students research famous mathematicians or other people who have made a contribution to mathematics education such as Euler, Descartes, Polya, Pythagorean van Hiele, or even Cuisenaire. Students could find out about experiences these mathematicians had, what they liked when they were elementary students, when they became interested in mathematics, or how long it took them to discover or create the mathematical theory or materials that made them famous. Students could prepare written reports on these people and share them with their classes. Then students might have a better understanding of what mathematicians do and why the students are mathematicians.

INTERVIEWING AS A MEANS OF WRITING

Fifth grade students researched famous mathematicians and then composed questions that would be appropriate to ask the mathematicians. In Figure 5.11, two students composed questions for Banneker. The questions about Major Andrew Ellicott and the kaleidoscope show that the students learned some things about Banneker.

The interview questions shown in Figure 5.12 were composed by two other students after reading about Sir Isaac Newton in an encyclopedia. An examination of the questions indicate that students may need instruction in composing probing questions for interviews. Teachers may model questions for an interview to help students gain an understanding of how mathematicians made their discoveries and how they used mathematics in their lives. Students can read newspaper and amagazine articles, or view films or videotapes about the person to be interviewed to help them prepare appropriate questions.

FIGURE 5.10. "I'M A MATHEMATICIAN"

I'm a mathematician because I
try hard. I never give up. I may
ask for help but I never give up.
I try to find a new way to do
hard things. Most things in life
have to do with math. If you
ever want to be someone you
have have to know math. I
may not like math, but I bet
you I'll try. This is why
I say I'm a mathematician.

FIGURE 5.11. FIFTH GRADE RESEARCH PROJECT

1. Why was you so attached to Mathematics?
2. What was mathematics like when you where a boy? Was it like the math today or was it diffrent?
3. How were the step's diffrent then ares in divison?
4. How was the assistant Major Andrew Ellicott like?
5. Were you a Math teacher ever?
6. How was your Math that you learn diffrent from today?
7. How Many years did you teach math? If you did?
8. Did you event the Kaleidoscopic?

FIGURE 5.12. QUESTIONS FOR AN INTERVIEW WITH ISAAC NEWTON

1. Who thought you mathematics.
2. How long did it take you to Learn mathematics.
3. Did you find any new ways To do mathematic divison problems.
4. Was math hard back when you were a child.
5. Did you have to check you divison problems.
6. Did you have Long divison problems.
7. Did you like math when you first began.
8. How old was you when you become a math scientist.
9. How long was you a math scientist.
10. Was you ever a math teacher.

COMPOSING INTERVIEW QUESTIONS

In addition to mock interviews with mathematicians, students can create interview questions to use with people living in their neighborhood who use mathematics in their daily lives, such as bank tellers, cashiers, salesmen, nurses, firefighters, and taxi drivers. To help students write good questions for their interviews, the teacher may invite another teacher or someone else who works in the school to come to the class to be interviewed. Students can ask questions of the person about ways he or she uses mathematics in their daily lives. The teacher can model and demonstrate good questioning techniques and work with students on the wording of their questions.

The teacher can demonstrate and discuss with students how to frame open-ended questions. Students may then role-play interviews with each other to help them learn how to write good questions that elicit appropriate responses. Students can ask probing questions of their subjects to help them gain an understanding of the opinions, experiences, and feelings of the person interviewed.

Once students have interviewed people in their own lives or conducted interviews with mathematicians of the past, they can be encouraged to write a summary of the information gained from the interview. Students can sequence the information gained and make a timeline of the person's life showing when where, and how they use mathematics in their daily lives. (Students may use the writing process described in Chapter 2 to help them write about mathematical experiences of the person interviewed.)

OBSERVING AND DESCRIBING

When students use appropriate descriptive language in describing the meanings of mathematical terms or describing relationships they are engaging in a type of communication categorized by Greenes, Schulman, and Spungin (1990) as "Observe and Describe." For example, when students are introduced to cubes (building blocks), they should be given time to play with them, to manipulate them, to build with them, and then to describe them. The teacher may have to ask questions to encourage kindergarten students to describe the cubes (blocks). By the time students are in the first grade, they should be encouraged to describe the cubes both orally and in writing. Certainly, from second grade and up, students should be encouraged to describe in writing various manipulatives, experiences, definitions, and relationships they have observed and experienced.

CREATING STORIES AND BOOKS ABOUT MATHEMATICS

Just as students create stories in language arts and reading, they can also develop an idea into a story or book about topics in mathematics. Students can sit in the "mathematician's chair," which is similar to the "author's chair" in language arts after tehy have written original stories and books with a mathematics content. (NCTM, 1989; Buschman, 1995.)

Students can write their own mathematical stories about familiar situations involving problems that can be solved with the aid of mathematics. This experience helps students learn to value mathematics and gain confidence in their mathematical abilities (Kliman and Richards, 1992). Figure 5.13 is a story that was composed and illustrated by two kindergarten students and word-processed by their teacher. Although the story is not about an everyday event, it is meaningful because it is about something the students know. Their class had just returned from a field trip to a zoo where they saw jaguars and birds. Once the class returned from the zoo, the teacher and the students discussed and described what they had seen and done while on their field trip. The students described situations they had observed that might contain ideas with a mathematical content. Students suggested possible mathematical situations. Then, working with a partner, students created and illustrated their own mathematical stories. Their teacher word-processed their stories and the students illustrated them.

Figure 5.14 shows a book entitled *The Number Book* written and illustrated by a first grade student. The student made drawings of objects to illustrate numbers 1 through 10. A 5-year-old illustrated a book about herself, her family, and numbers. Figure 5.15 shows one page from the story and illustrations.

Figure 5.16 was written and illustrated by a student. Her teacher, Patricia Hughes, stamped a clock face with no hands on several pages of paper for each student in her class. Students were studying time and when events usually happen. Students discussed events and times in their lives. Then Patricia's students wrote a story about themselves and sequenced events. Students described what they were doing at various times during the day and placed the minute and hour hand on the clock accompanying the story.

Tonya Pirlot's fifth grade students adopted first grade students as their buddies to serve as tutors in mathematics and reading. Some of Tonya's students made mathematics books for their buddies to explain mathematics concepts. Figure 5.17 shows a wizard pointing out that he will add the five apples on the tree and three apples on the ground. Next, the wizard is shown placing the three apples from the ground onto the tree. Finally, the result of the addition is shown.

(Text continues on page 116.)

FIGURE 5.13. WRITTEN AND ILLUSTRATED BY THE STUDENT, WORD-PROCESSED BY THE TEACHER

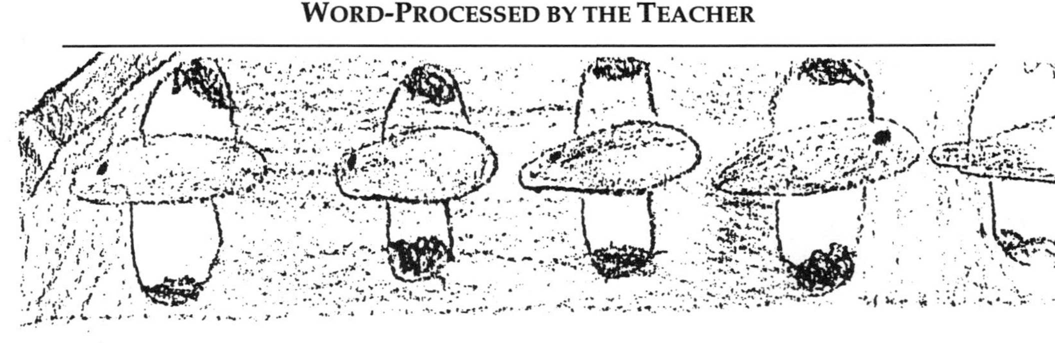

There are five birds flying and three jaguars come to play. Some were scared. Three birds flew away. How many are left? Mandy and Missy

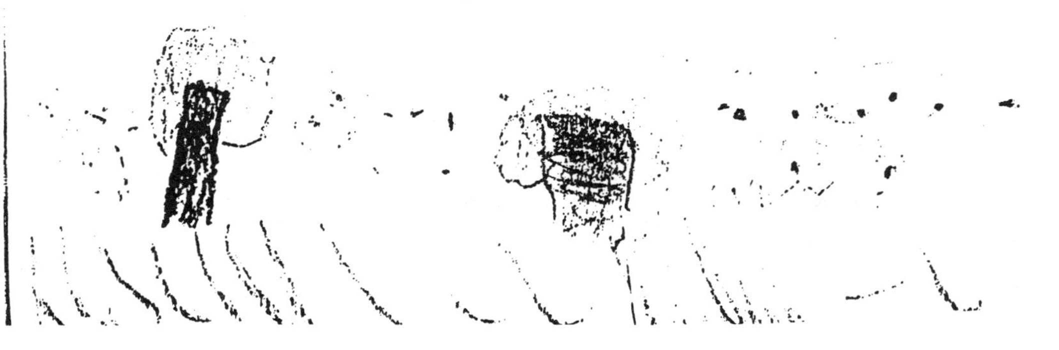

FIGURE 5.14. THE NUMBER BOOK

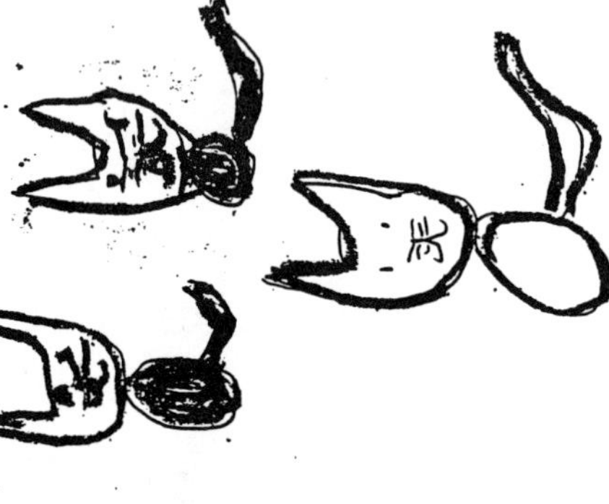

5 Five
6 Six
7 Seven

8 Eight.

9 Nine

10 ten
A B C D E
F G H I
J

1 2 3 4 5 6 7
8 9 10

FIGURE 5.15. NUMBERS AND THE FAMILY

FIGURE 5.16. WRITING ABOUT TIME

The first day
of school had
come. I ride
the bus to
school. The bus
gets to my bus stoo at
8:00. And it is 7:00.

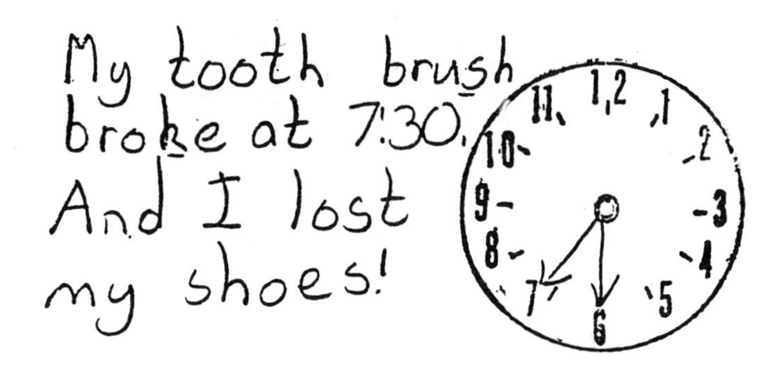
My tooth brush
broke at 7:30.
And I lost
my shoes!

Mom
where

I was at the bus
stop at 8:05.
I didn't know
if the bus
came or not.

At 8:30 school had started! My mom had to drive me to school.

FIGURE 5.17. FIFTH GRADER'S MATHEMATICS BOOK FOR HIS FIRST GRADE BUDDY

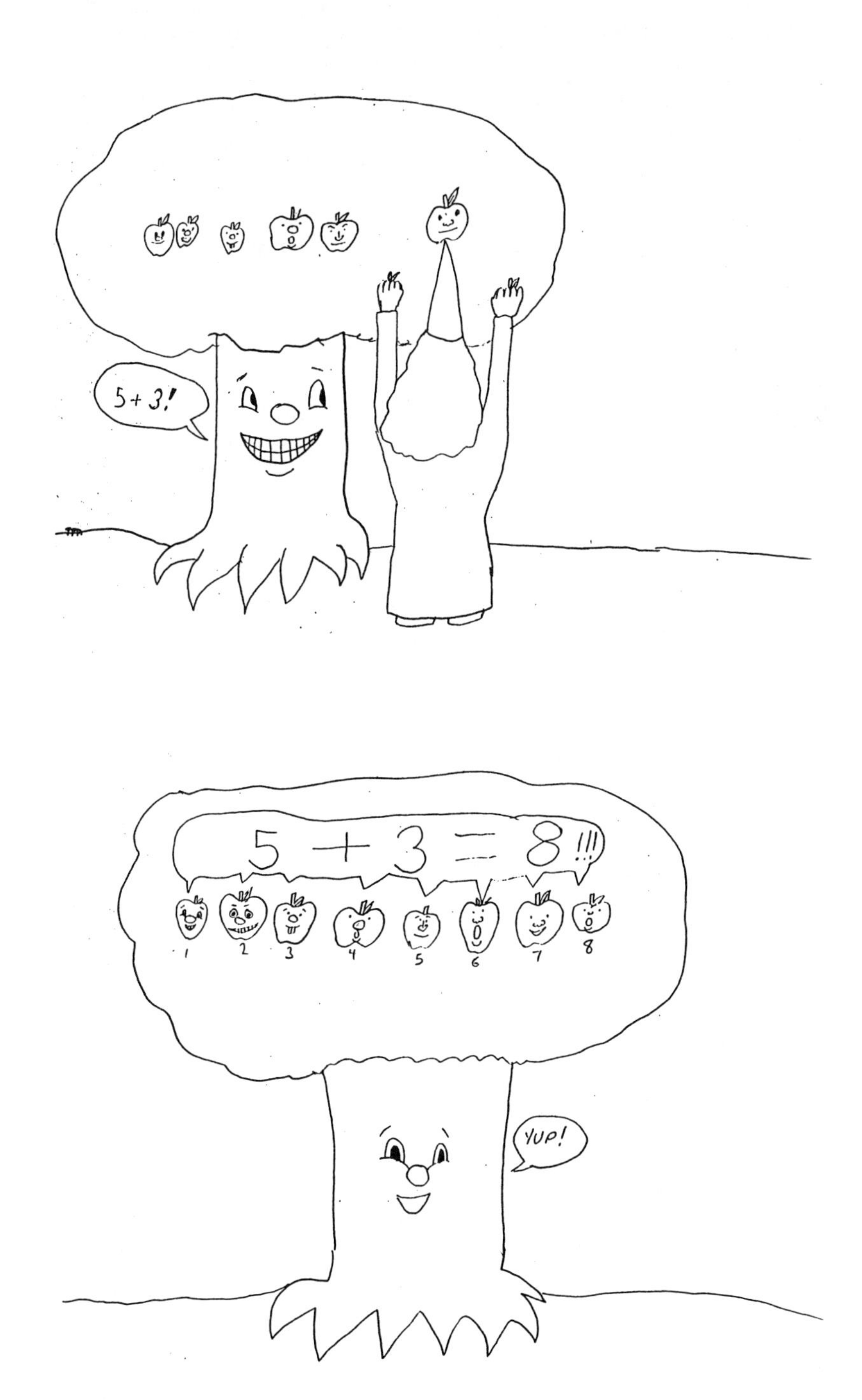
5 + 3!
5 + 3 = 8!!!
1
2
3
4
5
6
7
8
YUP!

Not only did the fifth grade students in Tonya's class help their first grade buddies by writing and illustrating books, they incorporated mathematical information into stories and modeled the operations for the young students. This activity benefits both of the buddies.

WRITING AND ILLUSTRATING PROBLEMS FROM TRADE BOOKS

There is a multitude of books written and published that are appropriate for mathematics. There are counting books, geometry concept books, books about time, measurement, money, sorting, classifying, numbers and logic (see Appendix B for a listing of books with a mathematics content).

Rebecca James had read *Five Little Monkeys Jumping on the Bed,* by Eileen Christelow, to her first grade class. After students discussed the story, Rebecca encouraged them to create and record a similar backwards counting story. Figure 5.18 shows some of the story written by a student who started at 10 and continued until "and they are dead." (The students in Rebecca's class are known as the "Kodiac Bears.")

Valerie Jones read *The Math Curse,* by Jon Scieszka and Lane Smith, to her fifth grade students. The story is about a student whose teacher tells him that everything in life is potentially a math problem, and sure enough, the student finds that he is under a "math curse." Figures 5.19 and 5.20 show stories by two of Valerie's students. *The Crazy Math Curse,* shown in Figure 5.19, takes an interesting twist in that the author includes problems and solutions to the problems throughout her story. Examination of the problems and solutions show that she has not allowed herself any time to eat breakfast. Her other situations are believable. The student has communicated her understanding of several concepts in mathematics.

The Math Problem, written by another student, is a creative open story (Fig. 5.20). The student-author does not present answers to any of the questions he has posed, but he certainly has incorporated mathematical information and relationships into his

(Text continues on page 120.)

FIGURE 5.18. BACKWARDS COUNTING STORY

10 LiTTle PeoPle
SiTTiNg ON a
KoDiaK Bear ONe
fell OFF aND BumeD
HiS HeaD.

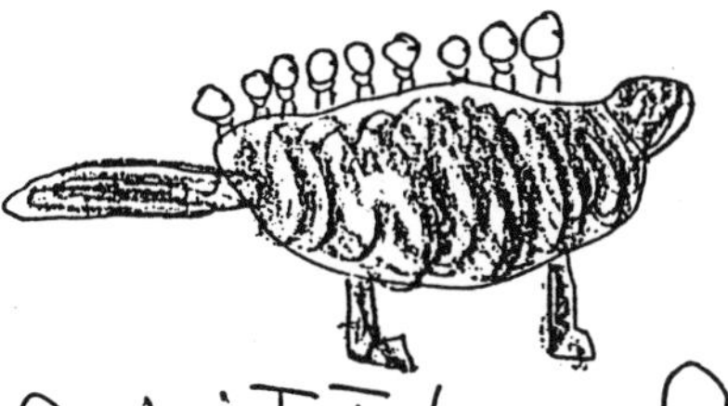

9 LiTTle PeoPle
SiTT aN a
KoDyAik Bear
ONe Fell OFF
aND BumPeD His
HeaD.

FIGURE 5.19. THE CRAZY MATH CURSE

During the summer, I would wake up at 7:23. I would go down to breakfast at 7:42. Then mom would drive me to the barn. I would get dressed before I went down to breakfast. How long did it take me to get dressed? If I got to the barn at 8:01, how long would it take me to get to the barn.

ANSWER: [It would take me 19 minutes to get dressed]

[It would take me 19 minutes to get to the barn]

When I get to the barn the horse teacher says "There are 2 pastures with 4 horses in them," "1 pasture with 7," "2 more with 12," "and 1 with 1." "So I need 6 horses bathed." If I need an equal amount of horses, how many will I need out of each pasture?

ANSWER: [We would need 1 from each pasture]

So we also have a barn full of horses. If we have 7 black, 12 bay, 3 spotted, 2 pallomio, 7 appoluso, 13 throghbread, 6 white, and 2 brown and black. How many horses do we have in all? How many horses do we have in the barn? Boy these math problems are going to my head!

ANSWER: [There are 52 horses in all]

[There are 12 horses in the barn]

So then I decide that I should take a math lunch break. So as I sit down to eat, a fly comes and drops on my food. Then I think how many flies are in this world. I think how many flies are in this world. I decide to call my mom and go home early today. After all I better start counting those flies.

1,2,3,4,5,6,7,8,9,10,11,12,13,14,15,16,17,18,19, 20,21...........

FIGURE 5.20. THE MATH PROBLEM

One day, I was coming home on the school bus and everything seemed like a math problem. How many people get on the school bus? How fast will I have to do my homework to still have time for supper and soccer practice?

Then I thought if I do my homework for 2 and a half hours, eat my supper in 15 minutes and drive to Fair Oaks in 10 minutes. I would wait 10 more minutes for practice to start. Then I would have to get home from school at 3:50.

The next day was worse. First, my friend wore a medal. How much medals would there be if everyone wore a medal but me, our class has 21 students? It got worse. Someone flipped their eye lids over. How many eye lids would be flipped if our entire class flipped both of their eye lids over? I tried to stop myself but I couldn't. Then someone in another class wore one earring on an ear. How many earrings would be worn, in my class, if everyone wore an earring on one ear?

a.23

b.22

c.21

That was not all. If I got home at 3:50 and do my homework for 2 hours, and play outside for 10 minutes, eat for 15 minutes, would I be late for P.S.R., which is at 6:30?

a.Yes

b.No

Soon I was dizzy from all the questions!

If there is 21 students in my class and then 5 went home how many students are in my class?

a.14

b.16

c.18

```
    The I realized that a big multiplication
problem could help! I tried this one:

              9,999,000
            x       102
             19,998,000
             00,000,000
          + 999,9000,000
          1,019,898,000

    That's how I broke the math curse!
```

story that are meaningful to him and has modeled familiar situations.

Each of the activities suggested for writing by students presents opportunities or students to communicate their understandings in mathematics.

REFERENCES

American Association for the Advancement of Science. *Science for All Americans: A Project 2061 Report.* Washington, DC: The American Association for the Advancement of Science (1988).

Atwell, Nancie. "Writing and reading literature from the inside out." *Language Arts* (1984), pp. 240–252.

Buschman, Larry. "Communicating in the language of mathematics." *Teaching Children Mathematics* (February, 1995), pp. 324–329.

Gordon, Christine, and Macinnis, Dorothy. "Using journals as a window on students' thinking in mathematics." *Language Arts* (January, 1993), pp. 37–43.

Greenes, Carole, Schulman, Linda, and Spungin, Rika. "Stimulating communication in mathematics." *Arithmetic Teacher* (October, 1992), pp. 78–82.

Helton, Sonia M. "I THIK THE CITANRE WILL HODER LASE: Journal keeping in mathematics class." *Teaching Children Mathematics* (February, 1995), pp. 336–340.

Kliman, Marlene, and Richards, Judith. "Writing, sharing, and discussing mathematics stories." *Arithmetic Teacher* (November, 1992), pp. 138–141.

Mumme, Judith, and Shepherd, Nancy. "Communication in mathematics." *Arithmetic Teacher* (September, 1990), pp. 18–22.

National Council of Teachers of Mathematics. *Curriculum and Evaluation Standards for School Mathematics.* Reston: VA: National Council of Teachers of Mathematics (1989) .

Perkins, David. *Smart Schools: From Training Memories to Educating Minds.* New York: The Free Press (1992).

Sanford, Betsy. "Writing reflectively." *Language Arts* (November, 1988), pp. 652–657.

Stix, Andi. "PIC-JOUR math: Pictorial journal writing in mathematics." *Arithmetic Teacher* (January, 1994), pp. 264–269.

Vacca, Richard T., and Vacca, Jo Anne. *Content Area Reading.* Glenview, IL: Scott, Foresman (1989).

Wilde, Sandra. "Learning to WRITE about mathematics." *Arithmetic Teacher* (February, 1991), pp. 38–43.

CHILDREN'S BOOKS

Christelow, Eileen. *Five Little Monkeys Jumping on the Bed.* New York: The Trumpet Club (1989).

Scieszka, Jon, and Smith, Lane. *Math Curse.* New York: Viking (1995).

6

Writing to Reason and to Make Mathematical Connections

The first part of this chapter describes two writing strategies that can be used to help students develop reasoning skills. The second part of the chapter addresses using writing to make mathematical connections with other areas of the curriculum and to everyday experiences of students.

WRITING TO REASON

Writing in mathematics can foster students' abilities to analyze, compare facts, and synthesize relevant material. When students write, they must organize their ideas into a synthesis of ideas that makes sense to themselves and to their readers. Writing is a recommended way for students to construct knowledge in that it helps them clarify thinking. The primary focus of school mathematics is to develop students' abilities to **reason** with mathematics and to apply mathematics to the solution of real world problems. This can be done through writing. (Emig, 1977; Fortescu, 1994; Mumme and Shepherd, 1990; Cangelosi, 1988; NCTM, 1989.)

Reasoning and critical thinking are developed as students sequence ideas and develop relationships in such activities as creating word problems, writing poetry about mathematics, or doing technical writing, such as creating books about how things work. In each of these activities, the student must organize and clarify the mathematical ideas in the stories, books, or poems they write. (Mayotte and Moore, 1992; Burns, 1995.)

Dialectical journals which provide focus questions and formats can be used to help students with thinking and reasoning skills. Thinking and reasoning skills can be developed by starting with simple thinking skills and moving to more complex ones. For example, when journals are first introduced, students can simply restate or paraphrase what they have read or discussed. Later, students can write interpretation entries where meaning is more personal. When students record applications they are responding at a higher level of thinking. When students analyze problems, create and justify solutions, and make applications, they are reasoning. (Edwards, 1991/1992.)

One fifth grade student, was asked by his teacher, David Barber, to select several items from a supermarket circular. He was to select items that are sold by the pound. Figure 6.1 shows the student's calculations for 3½ pounds of T-bone steak and 3½ pounds of New York strip steak. His explanation of how he figured the cost of 3½ pounds of steak is an example of his reasoning. David feels his students had a hard time writing their thoughts. David was, however, pleased that many of his students discovered that half of $0.99 should be considered as $0.50. The students who actually calculated one-half of $0.99 as $0.495, reasoned that they had to round up to $0.50. David believes that the students had a hard time explaining what they did in writing, because most of these students are not experienced writers. David thought most of his students did a good job of explaining what they did and he plans to do similar reasoning assignments.

PROBLEM SOLUTION JOURNALS

The recording of analyses, justifications, and applications in problem-solution journals helps develop reasoning skills. Edwards (1991/1992) presented the following example of a problem-solution journal with a math text:

What it says	What it means (draw a picture of the problem)	Operations	Solutions
(Students record problem here)			

FIGURE 6.1. PROBLEM SOLUTION USING A SUPERMARKET CIRCULAR

Select several items from a supermarket circular, from a newspaper, that is sold by the pound. List the items. Explain how you figure the cost of 3 1/2 pounds of each item.

T-Bone Steak 399 x 3 $11.97 + 2.00 = $13.97
N.Y. strip steak $4.99 x 3 $14.97 + 2.50 = $14.49
cHicken
SpLit FRyer Breast

we toke How much one pound costs and Multiplyyx 3 then take your ansewer and added with half of what you u strated with. We toke like 50 and Half of it would be a quater.

Edwards believes that the problem-solution journal requires students to go beyond inference and application (p. 314). The responses are at the synthesis-level of thinking. There is a developmental sequence in the growth of thinking and reasoning. Reasoning moves form the concrete to the abstract. Teachers must keep this in mind when they plan writing activities in mathematics. (Olson, 1984.)

WORD WEBS

A thought-organization and clarification strategy recommended by McGehe (1991) is a word web. A key word such as "square" is written in the center of the board. A circle is drawn around the word or concept. Students respond by writing what they know about a square. After students have brainstormed and written ideas about squares, the responses are written in categories by the teacher around the word on the chalkboard. The teacher does not reveal the categories to students. As students respond, hopefully they will figure out the categories of responses selected by the teacher and will suggest where to place additional responses. An example of a word web for a square with fourth grade students' responses is shown:

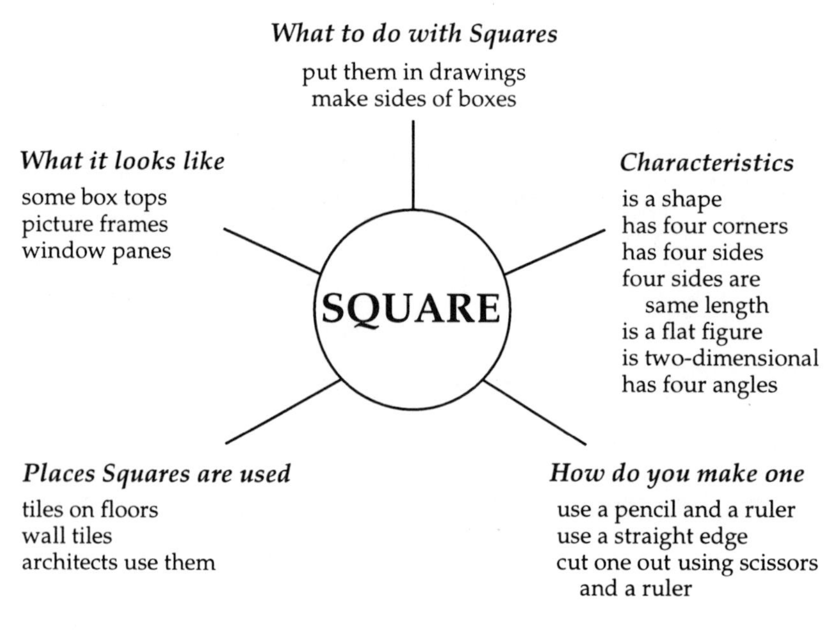

Once word webs are completed, students review how they have categorized their ideas. Are all responses placed in the appropriate categories? Are there other categories that we could add? Why was each response placed in a particular category? What have we learned about the [square] that we didn't know before? Are there other ideas to be placed on our word web? Are all recorded responses correct? How can we verify that all information is correct? The discussion following brainstorming and completing the word web is important in that it helps students focus on their reasoning and decide if they have made correct responses.

Writing explanations of reasoning or thinking strategies is appropriate for students in grades K–6 and students can write supporting arguments in journals. Writing is an appropriate strategy for the development of reasoning in mathematics.

WRITING TO MAKE MATHEMATICAL CONNECTIONS

One of the most widely accepted ways of making connections between mathematics and other areas of the curriculum is through the use of children's literature. Quality children's literature has become an important vehicle for integrating learning experience. The most effective literature in promoting true integration is using literature not explicitly designed for understanding mathematics (Smith, 1995). Many of the articles and books written about connecting mathematics and children's literature have used literature written for the purpose of directly exploring concepts in mathematics.

When teachers share literature regularly with their students they have a natural, meaningful context for encouraging students to write about mathematics. (Lewis, Long, and Mackay, 1993; Kliman, 1995.) Mathematics activities carried out with a fourth grade class which was reading *Gulliver's Travels* (Swift, 1983) was described by Kliman and Richards (1992). Students explored, through writing, drawing, and mathematics, the lands Gulliver visited. Students investigated places, characters, and events, and gather mathematical information they incorporate into their own writing.

A second grade student created the word problem shown in

Figure 6.2. After reading *Cinderella,* by Marcia Brown, the student created a story with mathematical content to accompany a story that contained none. To some extent, the story in her problem supports Smith's contention that "using literature not explicitly designed for understanding mathematics" is a way to integrate mathematics into the curriculum. *Cinderella* is a story that does not contain specific mathematics content, but the second grader used the story line to trigger mathematical ideas.

Figure 6.3 shows an original problem created by three students who worked together after reading *Caps for Sale* (Slobodkina, 1986). They wrote a mathematics problem about animals (there are monkeys in *Caps for Sale*) which is also shown in Figure 6.3.

Another student did not create a mathematics problem after hearing *Math Curse* (Scieszka and Smith, 1995). Instead, the book triggered an idea for the story she created (Figure 6.4). She used her real life experiences and creativity to generate the story.

LANGUAGE ARTS: POETRY

One teacher read the poem "Smart" (Silverstein, 1974) aloud to her class. She then discussed the poem with the students and had them sequence the poem's trading events. Figure 6.5 shows one student's sequencing of the trades described in "Smart" and her response to the question, "Was the boy in the [poem] smart or foolish?," and the direction to "Tell why" (you think so).

FIGURE 6.2. CREATING A STORY WITH MATHEMATICAL CONTENT TO ACCOMPANY ONE THAT CONTAINED NONE

FIGURE 6.3. ORIGINAL PROBLEMS WRITTEN AFTER READING CAPS FOR SALE

Caps for Sale

Esphyr Slobodkina

Once, there was a peddler that sold caps. He had 17 caps. One day He sat down to rest, and he wock up and he only had 1 hat. How many hats were gone.?

One day, the circus was coming to town. I Love Looking at the animals. There was one horse, one bear, two Lions, one elephant, two dogs, one bear, and three gorillas. How many animals were there at the circus in all?

FIGURE 6.4. STORY TRIGGERED BY READING MATH CURSE

The Horrible Math Week!

Is math your interest? well it sure isn't mine. I would just like to tell you about the HORRIBLE MATH WEEK! Well it all started when Mrs. Jones had said "class please think of some math problems for homework tonight". That is when it hit me smack in the head and it just seem to stick with me. That night thought to myself "every night I subtract 1 pair of jeans, 1 shirt, 2 socks, and 2 shoes". So the next day I thought of some more math problems. Then I said to myself "every morning I add 2 bows q pair of pants 2 shoes 2 socks 5 pencils and 1 bookbag". Here are my questions, What did I add that I didn't think of last night? How much is it all together?

The next day at school I added 2 erasers in my bookbag. Now how much is that?, including all of the other stuff. That night after school I had this dream about this math hall with problems everywhere. Suddenly I fell into this hole. After the dream I was free of that horrible math spell! Then Mrs. Jones said the next day!

"HOW MANY DAYS ARE IN A CENTURY"

AAAAAAAAAAAAAAAAAH!

FIGURE 6.5. RESPONSE TO QUESTIONS ABOUT SMART

1. The boy Started with a dollar. = $1.00

2. He traded for 2, quarters, = $.50

3. Then he traded for 3, dimes = $.30

4. Next he traded for, 4, nickels = $.20

5. Finally he traded for, 5 pennies = $.05

Was the boy in the story smart or foolish? Tell why.

I think he was very foolish because evey time he got money he would lose more money.

MATHEMATICS AND EVERYDAY EXPERIENCES

Mathematically literate students "view mathematics as a way of looking at their environment that aids understanding . . ." (Welchman-Tischler, 1992, p. 12). This means that they are able to see how mathematics is used in everything they do in everyday experiences. One fourth grade student tried to list 10 things he did each day that involved mathematics (Fig. 6.6). Another student completed the sample shown in Figure 6.7, describing ways she uses mathematics. Everything she describes involved time except, number 2 which is about money for ice cream.

There are many ways to use writing to connect mathematics and other content areas or mathematics and everyday experiences. Only a few have been described here, but the possibilities are endless.

(Text continues on page 138.)

FIGURE 6.5. "LIST 10 THINGS YOU DO EACH DAY INVOLVING MATHEMATICS"—I

Make a list of 10 things you do each day that involves mathematics (time, money, etc.).

When I walke up In the morning I lookat my alarm clock then I talke a bath then I get dressed. And I look at my watch to seewhen to leave. And my mom drives me to school And I go to the cafatirea to have breakfast. And I get my ticket to pay for my breakfast. A if I am still not done with my breakfast then I Look at the clock about every five minntes. A that then I go to my classroom and I Sharpen my Pencil befour eight O'clock and I always make shure that I sharpen my pencile befour eight O'clock. Then when the bell rings, we have a minute of quite reflection.

FIGURE 6.7. "LIST 10 THINGS YOU DO EACH DAY INVOLVING MATHEMATICS"—II

1. Waick up at | 6:30
2. Get my Ice crem money | ¢50
3. I look at the Date | 11/14/95
4. I go to the bus at | 7:10
5. we say the pledge at | 8:00
6. Do math at | 10:30
7. I go to Esol at | 11:00
8. Then we go to Lunch at | 12:10
9. Then we have silent reading time at | 2:00
10. The we go home at | 2:29

REFERENCES

Burns, Marilyn. "Writing in math class, absolutely." *Instructor* (April, 1995), pp. 40–47.

Cangelosi, James S. "Language activities that promote awareness of mathematics." *Arithmetic Teacher* (December, 1988), pp. 6–9.

Edwards, Phyllis R. "Using dialectical journals to teach thinking skills." *Journal of Reading* (December 1991/January, 1992), pp. 312–316.

Emig, Janet. "Writing as a mode of learning." *College Composition and Communication* (1977), pp. 122–128.

Fortescue, Chelsea M. "Using oral and written language to increase understanding of math concepts." *Language Arts* (December, 1994), pp. 576–580.

Kliman, Marlene. "Integrating mathematics and literature in the elementary classroom." *Arithmetic Teacher* (February, 1993), pp. 318–321.

Kliman, Marlene, and Richards, Judith. "Writing, sharing, and discussing mathematics stories." *Arithmetic Teacher* (November, 1992), pp. 132–141.

Lewis, Barbara A., Long, Roberta, and Mackay, Martha. "Fostering communication in mathematics using children's literature." *Arithmetic Teacher* (April, 1998), pp. 470–473.

Mayotte, Gail, and Moore, Christine. "Writing, thinking and math." *Teaching K–8* (January, 1992), pp. 49–51.

McGehe, Carol A. "Mathematics the write way." *Instructor* (April, 1991), pp. 36–38.

Mumme, Judith, and Shepard, Nancy. "Communication in mathematics." *Arithmetic Teacher* (September, 1990), pp. 18–22.

National Council of Teachers of Mathematics. *Curriculum and Evaluation Standards for School Mathematics.* Reston, VA: National Council of Teachers of Mathematics (1989).

Smith, Jacquelin. "A different angle for integrating mathematics." *Teaching Children Mathematics* (January, 1995), pp. 288–293.

Welchman-Tischler, Rosamond. "Making mathematical connections." *Arithmetic Teacher* (May, 1992), pp. 288–293.

CHILDREN'S BOOKS

Brown, Marcia. *Cinderella*. New York: Macmillan (1954).

Scieszka, Jon, and Smith, Lane. Math Curse. New York: Viking (1995).

Silverstein, Shel. "Smart" in *Where the Sidewalk Ends*. New York: HarperCollins Publisher (1974).

Slobodkina, Esphyr. *Caps for Sale*. New York: Harper Trophy (1986).

Swift, Jonathan. *Gulliver's Travels*. New York: William Morrow Co. (1983).

7

Writing to Assess Learning in Mathematics

Instruction in mathematics should reflect higher-order thinking in the assessment process. In 1995, NCTM produced its *Assessment Standards for School Mathematics* to expand on and complement its *Evaluation Standards* (p. 1). The *Assessment Standards* document defines assessment as "the process of gathering evidence about a student's knowledge of, ability to use, and disposition toward mathematics, and of making inferences from that evidence for a variety of purposes" (p. 3). The *Assessment Standards* are based on the assumptions that "all students are capable of learning mathematics and their learning can be assessed" (p. 1). The *Assessment Standards* reflects a shift toward evaluation by teachers using several sources and "toward using concepts and procedures to solve problems" (p. 3).

Journal writing, either free-writing or entries written in response to prompts, is a recommended way to gather evidence about students' knowledge and the use of mathematics.

Writing can afford the teacher a unique opportunity to assess students' understanding and fluency of the concepts about which they are writing. The teacher can assess whether students are learning concepts by examining the writing of students in mathematics. (Sosenke, 1994; McIntosh, 1991.) Writing can serve as a diagnostic tool by which the teacher is able to ascertain a student's understanding of concepts. Through writings in mathematics, a teacher can gain insights into students' thinking and can look at how they arrive at their conclusions (Wilde, 1991; Mumme and Sheperd, 1990). The teacher gains insights into students conceptions and misconceptions through their writing.

The Toronto Board of Education mandated the development of standards for student achievement in mathematics, including assessment (Clark, 1992). Data gathered for the assessment of student achievement included students' written work. The combined use of writing and cooperative learning can be used to enhance assessment techniques (Artz, 1994). In this chapter, various kinds of writing by students are examined for the purpose of assessment.

ASSESSING CONCEPTS

A first-grade student "wrote" and illustrated his understanding of the numbers 1 through 10. Figure 7.1 shows his explanation. When the teacher examines his writing, she can be fairly certain that he understands the number of objects to accompany counting from 1 to 10. She needs to further check the student's concepts to be sure that he understands that there are several combinations that have sums equal to 10.

Figure 7.2 was written by a Kindergarten student whose teacher asked that she trace around a group of attribute shapes and to draw a line between or around alike shapes. The Kindergarten student was successful at tracing the shapes and sorting them. Each shape was a different color. When the student traced the shapes, she also colored them. It is, therefore unclear, whether she sorted by shape as the teacher had asked or whether she sorted by color. The teacher must, in this case, plan other writing activities with attribute shapes that are all the same color to see if students are sorting by shape or color.

Figure 7.3 shows a writing sample from a second grade student. To check the student's understanding of the value of coins and to see if she could make equivalent amounts of money using nickels and dimes, her teacher placed school supplies (scissors, pencils, a box of crayons, and a bottle of glue) with price tags attached to each at a center. Students were to make a drawing of the school supplies and the price tags. They were to draw the correct number of nickels and/or dimes needed to make the purchase of each item. She has drawn the correct number of coins to purchase each item.

(Text continues on page 148.)

FIGURE 7.1. ASSESSING UNDERSTANDING IN YOUNG CHILDREN

FIGURE 7.2. SORTING BY SHAPES

FIGURE 7.3. UNDERSTANDING THE VALUE OF COINS

Student responses to prompts can be used to assess their understanding of mathematical concepts. Students can explain, in writing, their "reasoning and evaluate whether a problem was too easy, just right or too hard" after they have , for example, compared ⅔ and ¾ (Burns, 1995). (A list of prompts for writing in mathematics is found in Appendix C.)

A second grade student responded to the prompt, "How do you measure?" by describing measuring with rulers, yardsticks, and a thermometer (Fig. 7.4). The teacher might follow the student's response with other prompts that ask her to describe what she knows about measuring capacity, perimeter, area, and time. For example, the teacher might give the prompt, "How do you measure the amount a container will hold?"

Figure 7.5 shows the response of a third grade student to the prompt, "How do you measure liquid?" Susan Fields, who was working on a unit on measurement, gave this prompt to her students along with, "How do you measure weight?"; "What is measuring?"; "What are centimeters?"; and "How do you use a ruler?" Figure 7.6 shows the students' responses. An examination of the responses permits Susan to examine exactly what each student in her class knows about measurement, allows her to learn where students are in their development of reasoning, and helps her make plans for individual, small group, and total group instruction.

A 9-year-old student responded to the prompts "What is money?"; Why do we use it?"; "What is time?"; and "How do we use it?" (Fig. 7.7). An examination of her responses tells her teacher a great deal about her level of reasoning and suggests that the teacher has some work to do to help students continue developing concepts of measurement.

Students can respond to prompts, write to answer questions or to explain what they know about something. Their responses can be examined for the purpose of assessing what they understand about the concepts being addressed. Figure 7.8 shows one student's written journal response to the prompt, "What I Know About Telling Time." This student's teacher can tell from her writing that she can identify the hour and minute hands on a clock, but cannot determine if the student can "tell" time.

(Text continues on page 154.)

FIGURE 7.4. "HOW DO YOU MEASURE?"

Wen you measure
a peic of wood you
need a yuler wen
you measare a Big
peic of wood you
need a yeard
stek. A termatr
measeres the
tenpature.

FIGURE 7.5. "HOW DO YOU MEASURE LIQUID?"

How do you mesure liquid?

You mostly mesure liquid by using cups, pints, quarts and gallons. Cups are used to mesure little liquids, pints are used to mesure middle sized liquids, quarts are used to mesure big liquids and gallons are used to mesure real big liquids, For an example cups lock like this  pints look like this

pints

FIGURE 7.6. "HOW DO YOU USE A RULER?"

5-31-97

How To Use A Ruler

Inches (In)

I will start with the customary system. The inch is the smallest measurement in the customary system. I will show you somthing That shows 1 inch.

1/4 1/2 3/4 1 inch
12 inches = 1 Foot

Feet (FT)

Feet are after inches in the customary measurement system. I will show you a foot on the back of this page. 3 feet = 1 yard

Yards (Yd)

Yards are bigger then inches or feet, but they are not the biggest measurment. Even though it is not the biggest measurement, it is so big I can't show it to you on paper!
1,000 (Yd) = 1 Mile

FIGURE 7.7. "WHAT IS MONEY?" "HOW DO WE USE IT?" "WHAT IS TIME?" "HOW DO WE MEASURE IT?"

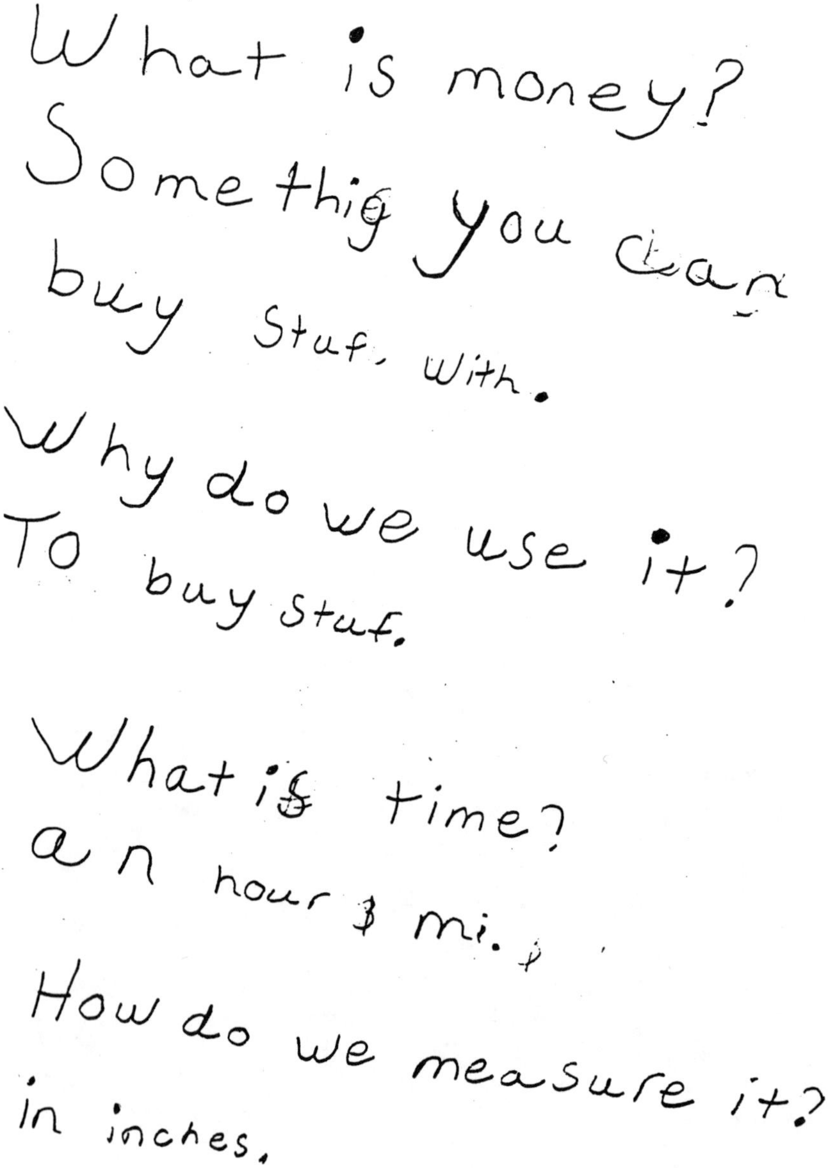

What is money?
Somethig you can buy stuf with.
Why do we use it?
To buy stuf.

What is time?
an hour & mi.
How do we measure it?
in inches.

FIGURE 7.8. "WHAT I KNOW ABOUT TELLING TIME"

What I Know about telling time.
The clock is very importent.
It is how we tell time. This is how you use a clock. The little hand is called a hour hand. The big hand is called the minute hand.

FIGURE 7.9. IDEAS ABOUT CIRCLES

I am going to be talking about circles. A circle is a shape. It is sorda like a ball. A circle doesn't have corners. You can make alot of stuff with circles, like a spinning weel. Evrywere you go you will see a circle. Circles are evrywer Circle can be any colors. My favrite color as a circle is red. Circles look like planets. They can be made by you. All you need is pappee and a pencil.

A fourth grade student described what he knows about circles in Figure 7.9 (shown on prior page). He wrote the response when his teacher asked students to write ideas about circles. This was done as a part of a brainstorming of ideas to be included in a word web. This student seems to know several things about circles—they are shapes and they have no corners. The other information he describes and his illustration appear to indicate that that he knows circles are plane figures ("All you need to make one is pap(p)er and pencil"), but it is difficult to tell if he confuses circles and spheres when he writes "Circles look like planets." To clarify whether this student's concept of circles is correctly formed, the teacher might interview him and ask questions that reflect more of what he understands about circles.

Figure 7.10 is a third grade student's answer to the prompts, "Describe how 18 cookies can be shared equally by 5 students;" "Describe how you figured out how many each student will get." These prompts were given to students after reading *The Doorbell Rang*, by Pat Hutchins. The teacher wanted to see if her students could explain division with a remainder. The third grader's explanation shows that she understands division as partitioning. She did not write that she would divide to share the cookies, but she would give them out in a dealing-out manner. The response shows she does understand the concept of division, but leaves the teacher with questions about the student's connection to the specific mathematics concept..

ASSESSING PROBLEM-SOLVING

Many examples of original word problems created by students are included in this chapter. Some show the creativity of the students who wrote them. Some show that students do not ask realistic questions with their story problems, but ask typical textbook-type questions using clue words. Still other samples show that some students have neither an accurate understanding of what makes a word or story problem nor of what they need to include when they write a problem.

Figure 7.11 shows two "word problems" written by a 10-year-old student who attended a summer program I directed in 1993. I asked students to write one or two word problems similar to

FIGURE 7.10. "DESCRIBE HOW 18 COOKIES CAN BE SHARED EQUALLY . . ."

Describe how 18 cookies can be shared equally by 5 students. Describe how you figured out how many each student will get.

I would count how many Cookies I had. Then I would pass them out until I was done and if there was any remanders I would not give them out

FIGURE 7.11. WORD PROBLEMS SHOWING A STUDENT'S LACK OF UNDERSTANDING OF INFORMATION NEEDED IN WORD PROBLEMS

Miss Swamp camein 24 times
and she was at home 24 tomes
How Many are left

The kids were in room
207, And everyday was
The same room 207. How
Many are in all.

those they had experienced during the school year. This sample shows that the student does not understand what kind of information must be included in a word or story problem. It is evident that the student is familiar with two questions that are typically asked in textbook word problems, "How many are left?" and "How many in all?"—both include clue words.

The two word problems shown in Figure 7.12 were also written by a 10-year-old who attended a summer program. This student used the same question for an addition problem and a multiplication problem, "How many . . . in all?" This student did include enough information for the reader to be able to find an answer to the questions asked. An examination of the information in the problems written by this student alerts the teacher that he may need additional experiences for developing more realistic problem situations and in writing more realistic questions that do not use on word clues such as "in all."

FIGURE 7.12. WORD PROBLEMS USING WORD CLUES

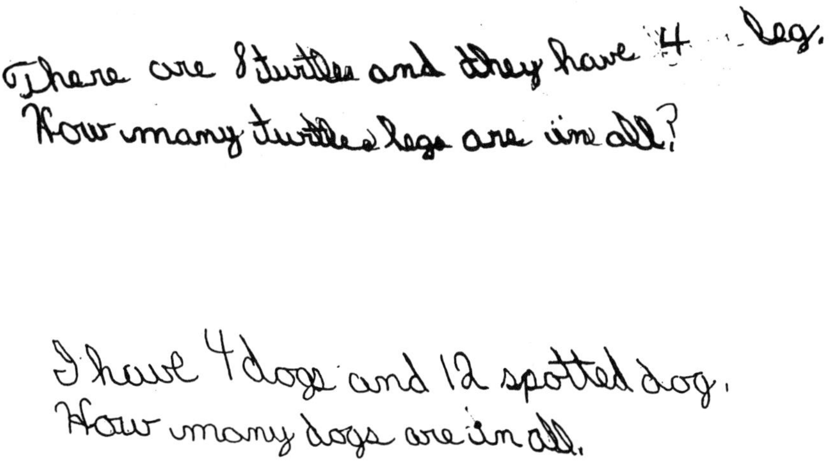

Thare are 8 turtles and they have 4 leg.
How many turtles legs are in all?

I have 4 dogs and 12 spotted dog.
How many dogs are in all.

Figure 7.13 shows a word problem written and illustrated by a fifth grade student in her regular mathematics class. Her problem contains all necessary information for the reader to answer

FIGURE 7.13. WORD PROBLEM CONTAINING ALL NECESSARY INFORMATION

MEL OWNS A STORE. HIS FRUITS ARE ON SALE FOR: 39¢. LAST WEEK MEL'S FRUITS WERE ON SALE FOR: 85¢ MORE. WHAT WERE THE PRICE OF THE FRUIT LAST WEEK? $1.24

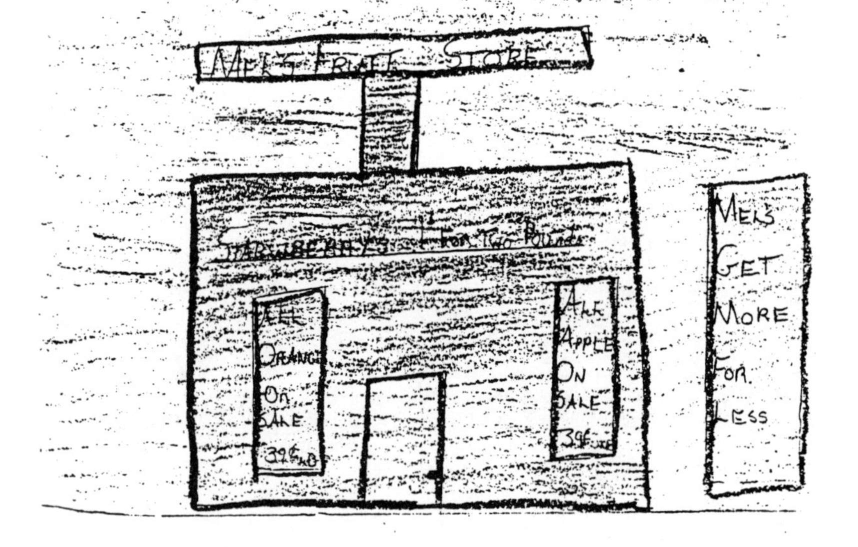

the question she posed. The question is not a typical "How many in all?" or "How many are left?" question.

The word problem shown in Figure 7.14, is a realistic problem that comes from the student's own experience. The student listed 14 scores for "Amy" and asks the reader to "find Amy's average." The student-author gave a hint about how to find the average and then indicated that Amy's average was 85.3, an incorrect response. The average of the scores presented in the problem is 78. The student's teacher should check to determine whether the student made an addition or division error. The problem does show that the student knows how to create a problem situation.

The word problem shown in Figure 7.15 looks good because of the illustration. However, an examination of the word problem itself and the solution shows a great deal about the student-author's problem posing and solving. The problem posed is an interesting one, but the solution is faulty. An examination of the

FIGURE 7.14. REALISTIC WORD PROBLEM FROM STUDENT'S EXPERIENCE

Amy has 14 score for math.
(100, 65, 81, 17, 93, 100, 73, 89, 94, 100, 100, 61, 79 & 41.
Use the scores above to find Amy average
What is Amy score for Math?
How did you find her average
Hint add all 14 # & divide by 14.
Her average 85.3

FIGURE 7.15. WORD PROBLEM WITH FAULTY SOLUTION

My Math Word Problem

If it takes 2 hours to get to Alabama going 65 mph. How long will it take to get to Alabama going 55 mph? 3 hours 11 min.

Answer: 2 hours ÷ by 55 mph. = 27.5. Take 60 min from the 75 min (75) and add it to the 2 hours, which

$$\begin{array}{r} 75 \\ -60 \\ \hline 15 \end{array}$$

makes 3 hours plus the 15 min left over. So the answer is 3 hours and 11 min.

student's explanation shows that he has no real understanding of how to calculate rate and time. When his teacher examines the problem and solution, she should be aware of the problems the student is having.

Six students worked together to create the story of "Laura Jean, the Yard Sale Queen" shown in Figure 7.16. This group-created problem and question are both creative and realistic.

FIGURE 7.16. CREATIVE AND REALISTIC WORD PROBLEM

Laura Jean The Yard-Sale Queen

Laura Jean was trying to clean out the basement for Lulu's bed. She was going to have a yard sale. And she sold a Kazoo to Professor Farnsworth for $5.00. She sold a boa and a pair of earing for $2.50. Then she sold a two china dogs to Big Stan for $2.00. How much does she earn?

The three word problems shown in Figure 7.17 show that the student-author understands problems in which multiplication is an appropriate operation to use to find the solution. His first and third questions involve the multiplication of three numbers, a concept that his teacher explored with students.

FIGURE 7.17. WORD PROBLEMS DEMONSTRATING THAT THE STUDENT UNDERSTANDS MULTIPLICATION

① Ther are three aligators if each
aligator has four claws and each
aligator has four legs how many
claws are there all to gether?

② there are three Sea mansters
and each Sea monster has four
Wings. how many wings are there
all together?

③ there are six pirates each
pirate eats 7 fish aday
six day past how many
fish did the pirates eat all together?

ASSESSING COMPUTATION

One teacher presented students with two "problems," one with the correct answer and the other with an incorrect answer. Students were asked to identify the incorrect answer, explain why it is incorrect, and to correct the answer. Figure 7.18 shows one student's explanation of which problem was incorrect. The student whose explanation appears in Figure 7.19 corrected his own mistake and described what he had done incorrectly. This student's teacher can see from his writing that he can "follow steps" in dividing. The sample does not assess whether he actually understands the concept of division of large numbers. Both of these samples show isolated algorithms. For their teachers to assess their true understandings of adding or dividing, the teachers need to have students solve problems in the context of real problem situations.

ASSESSING ATTITUDES

In "What I Like About Math" (Fig. 7.20), the student-author states that he likes math because he does well in mathematics. It seems from his explanation that he likes all strands of math and thinks math is "good for you." He also thinks "you will need math when you get older—like in high school." This student's teacher can assume he enjoys math and is willing to tackle any problems presented.

The explanation of probability shown in Figure 7.21 indicates that the student who wrote it has an emerging sense of probability and is comfortable in stating a real life example about a roller coaster. This could tell the teacher that the time is right for more exploration of probability.

The student writings shown throughout this chapter are examples of how writing can be used as a way to assess what students understand. The writings of students in mathematics can also be a guide for the teacher in planning instruction. Student thought processes, perceptions, and attitudes about mathematics can be revealed in their writing in mathematics. Students learn about their own thinking when they write in mathematics. Their

(Text continues on page 168.)

FIGURE 7.18. EXPLAINING WHY AN ANSWER IS INCORRECT

Which problem is incorrect?
Explain why?
What must you do to correct the problem?
Work out the problem?

1. 72 − 47 = 25

Problem two is incorrect.
The part of the ones is right but the part with tens in not right. Joanne bid not sudtract that is why it is not right.

You most sudtrdct the tens side to make it right.

54 − 16 = 38

FIGURE 7.19. STUDENT EXPLAINS HIS OWN MISTAKE

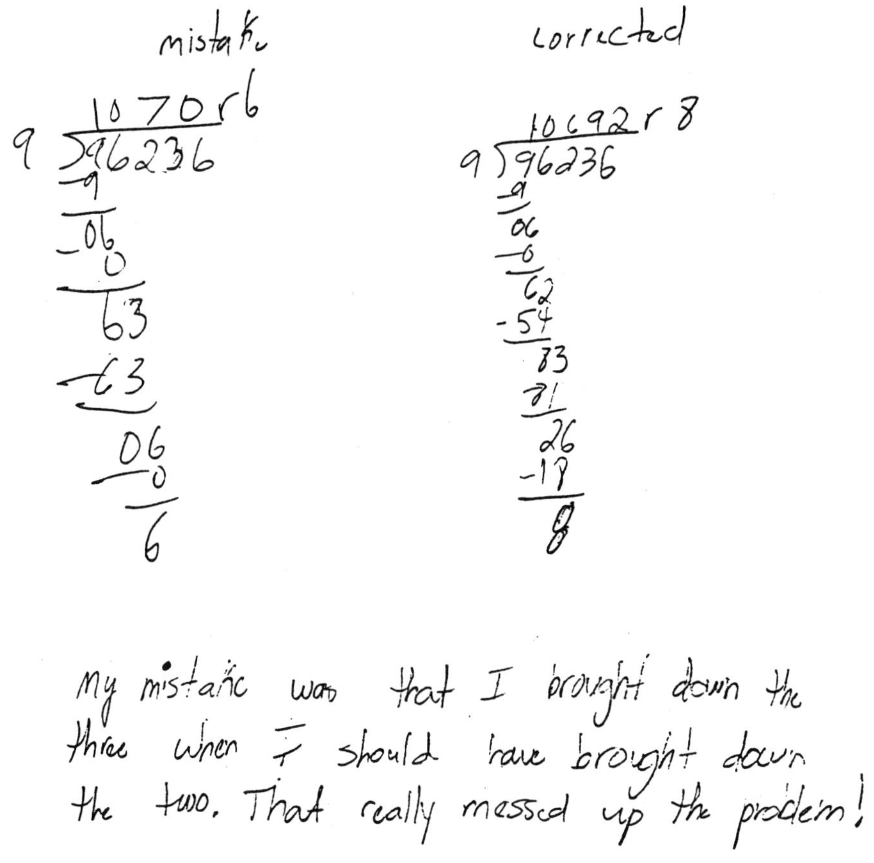

mistake

1070 r6
9 ⟌ 96236
-9
06
-0
63
-63
06
-0
6

corrected

10692 r 8
9 ⟌ 96236
-9
06
-0
62
-54
83
-81
26
-18
8

My mistake was that I brought down the three when I should have brought down the two. That really messed up the problem!

FIGURE 7.20. "WHAT DO YOU LIKE ABOUT MATH?"

What I like about

math

I like math because of all the fun things we do here in math. Math is fun because i lik the liters and adding, adding i like adding because i all was get it right. subtracsun is one of my faverit subgeks, dividing is my favite subgekt because i like to work with hi numbers in school some time I get very thing right. math is very good for you because it teach is you about things at school, school is fun because math is fun because it is teaching you things that you proble didn't now. I like math because of the subtraction and multiplying. I can not wait for it to start. I like math because it is the last subject of the day. In math i like to use centimeter and kilometer. In math i like to use the grameses and other kind of math. Math is good for you because it can teach you how to do hard things in the math book. I think you should do math because it can make your grade hier and it can make you smart in school. math is some time boring but you will see that you need this subject when your are older and when you are in a hieher grade like in hiegh school you will need it in your life, math is fun

Adam
11/10/94
math
writing

FIGURE 7.21. PROBABILITY

What is a probability?

A probability is the chance of something. The more you have, the less probability, or chance, you have. I will give you some examples

When you get on the roller coaster, there is a 50 50 chance you will scream.

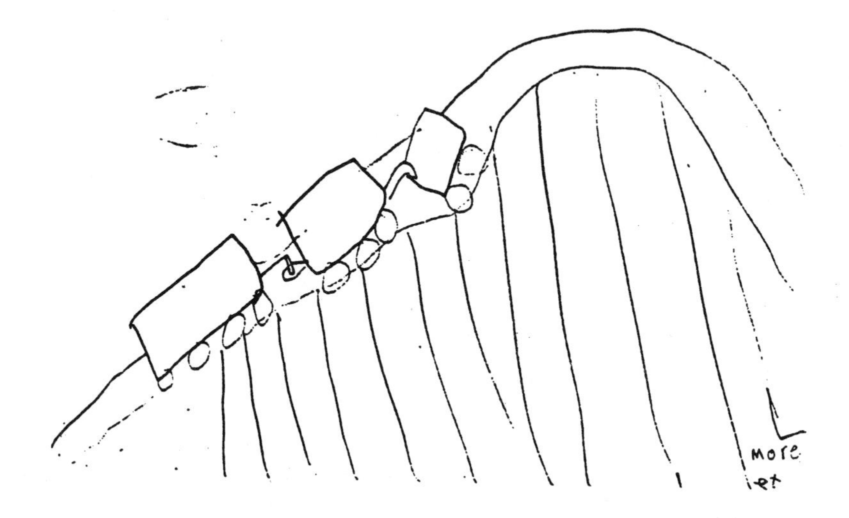

understandings and thoughts are clarified and as they become better at writing, whether it be explanations or creating story problems, students become empowered as mathematicians. Students learn to value mathematics and become aware of its use in everyday experiences. They are more likely to take charge of their own learning. Textbooks often include only problems with a limited focus which restricts students' abilities to develop flexibility in their problem-solving approaches. When students are allowed to develop their own problem-solving strategies, particularly through writing, a greater level of understanding will be achieved.

REFERENCES

Artzt, Alice F. "Integrating writing and cooperative learning in mathematics." *The Mathematics Teacher* (February, 1994), pp. 80–85.

Clark, John. "The Toronto Board of Education's benchmarks in mathematics." *Arithmetic Teacher* (February, 1992), pp. 51–55.

Gordon, Christine J., and Macinnis, Dorothy. "Using journals as a window on student thinking in mathematics." *Language Arts* (January, 1993), pp. 37–43.

McIntosh, Margaret E. "No time for writing in your class?" *The Mathematics Teacher* (September, 1991), pp. 423–433.

Mumme, Judith, and Shepard, Nancy. "Communication in mathematics. *Arithmetic Teacher* (September, 1990), pp. 18–22.

National Council of Teachers of Mathematics. *Curriculum and Evaluation Standards for School Mathematics.* Reston, VA: National Council of Teachers of Mathematics (1989).

Sosenke, Fanny. "Students as Textbook Authors." *Teaching Mathematics in the Middle* (September/October, 1994), pp. 108–111.

Stenmark, Jean Kerr, editor. *Mathematics Assessment. Myths , Models, Good Questions and Practical Suggestions.* Reston, VA: National Council of Teachers of Mathematics (1991).

Wilde, Sandra. "Learning to write about mathematics." *Arithmetic Teacher* (February, 1991), pp. 38–43.

CHILDREN'S BOOKS

Hutchins, Pat. *The Doorbell Rang.* New York: Mulberry Paperback Books (Greenwillow) (1986).

APPENDICES

APPENDIX A
CHILDREN'S BOOKS FOR PARALLEL STORIES

Aker, Suzanne. *What Comes in 2's, 3's and 4's?* New York: Simon & Schuster for Young Reading (1990).

Allen, Pamela. *Mr. Archimedes Bath.* New York: Crowell (1977).

Anno, Masaichiro, and Anno, Mitsumasa. *Anno's Mysterious Multiplying Jar.* New York: Philomel Books (1983).

Anno, Mitsumasa. *Anno's Counting Book.* New York: Crowell (1977).

Bang, Molly. *Ten, Nine, Eight.* New York: First Mulberry Edition (1991).

Carle, Eric. *The Grouchy Ladybug.* New York: Harper Trophy Books (1986).

Christelow, Eileen. *Five Little Monkeys Jumping on the Bed.* New York: The Trumpet Club (1989).

Clement, Rod. *Counting on Frank.* Milwaukee: Gareth Stevens (1991).

Dee, Ruby. *Two Ways to Count to Ten* (illustrated by Susan Meddaugh). New York: Henry Holt & Co. (1988).

Ehlert, Lois. *Fish Eyes: A Book You Can Count On.* San Diego: Harcourt Brace Jovanovich (1990).

Galdone, Paul. *Over in the Meadow.* New York: The Trumpet Club (1992).

Giganti, Paul, Jr. *How Many Snails?* (pictures by Donald Crews). New York: Greenwillow (1994).

Hamm, Diane Johnston. *How Many Feet in the Bed?* (illustrated by Kate Sally Palmer). New York: Simon & Schuster (1991).

Hoban, Lillian. *Arthur's Funny Money.* New York: Harper & Row (1981).

Hutchins, Pat. *The Doorbell Rang.* New York: Greenwillow Books (1986).

Mathews, Louise. *Bunches and Bunches of Bunnies* (pictures by Jeni Bassett). New York: Scholastic, Inc. (1978).

Mathis, Sharon Bell. *Hundred Penny Box* (illustrated by Leo & Diane Dillon). New York: Puffin (1986).

Reid, Margarette S. *The Button Box* (illustrated by Sarah Chamberlain). New York: Puffin Unicorn Books (1990).

Schwartz, David. *How Much Is A Million?* (pictures by Steven Kellogg). New York: Lothrop, Lee, & Shephard Books (1985).

Schwartz, David. *If You Made A Million.* New York: Scholastic (1989).

Slobodkina, Esphyr. *Caps for Sale.* New York: First Harper Trophy Edition (1987).

Tompert, Ann. *Grandfather Tang's Story* (illustrated by Robert Andrew Parker). New York: Crown Publishers (1990).

Viorst, Judith. *Alexander Who Used to Be Rich Last Sunday* (illustrated by Ray Cruz). New York: Atheneum (1978).

APPENDIX B
CHILDREN'S MATHEMATICS LITERATURE

Anno, Mitsumasa. *Anno's Counting Book.* New York: Thomas Y. Crowell Publishers (1977).

Anno, Mitsumasa. *Anno's Counting House.* New York: Philomel Books (1982).

Anno, Mitsumasa. *Anno's Math Games.* New York: Philomel Books (1987).

Anno, Mitsumasa. *Anno's Math Games II.* New York: Philomel Books (1989).

Birch, David. *The King's Chessboard.* New York: Dial Books for Young Readers (1988).

Brett, Jan. *The Twelve Days of Christmas.* New York: Dodd, Mead & Co. (1986).

Burningham, John. *Just Cats.* New York: Viking Press (1983).

Burns, Marilyn. *The I Hate Mathematics Book.* Boston: Little, Brown & Co. (1975).

Burns, Marilyn. *Math for Smarty Pants.* Boston: Little, Brown & Co. (1982).

Burns, Marilyn. *This Book Is About Time.* Boston: Little, Brown & Co. (1978).

Carle, Eric. *My Very First Book of Numbers.* New York: Crowell Junior Books (1985).

Crews, Donald. *Ten Black Dots.* New York: Greenwillow Books (1986).

deRegniers, Beatrice Schenk. *So Many Cats.* New York: Clarion Books (1985).

Eichenberg, Fritz. *Dancing in the Moon: Counting Rhymes*. New York: Harcourt, Brace, Jovanovich (1983).

Emberley, Rebecca. *Drawing with Numbers and Letters*. Boston: Little, Brown & Co. (1981).

Feelings, Muriel. *Moja Means One: A Swahili Counting Book*. New York: Dial Books for Young Readers (1971).

Friskey, Margaret. *Chicken Little, Count-to-Ten*. Chicago: Children's Press (1946).

Hoban, Tana. *Circles, Triangles, and Squares*. New York: Macmillan Publishing Company (1974).

Hoban, Tana. *Shapes and Things*. New York: Macmillan Publishing Co. (1970).

Hoban, Tana. *Shapes, Shapes, Shapes*. New York: Greenwillow Books (1986).

Hutchins, Pat. *One Hunter*. New York: Greenwillow Books (1982).

Kredenser, Gail. *One Dancing Drum*. New York: S.G. Phillips (1971).

Law, Felicia, and Chandler, Suzanne. *Mouse Count*. Milwaukee, Wis.: Gareth Stevens (1985).

LeSieg, Theo. *Ten Apples Up On Top*. New York: Beginner Books (1961).

McKissack, Patricia. *A Million Fish . . . More or Less*. New York: Alfred A. Knopf (1992).

McMillan, Bruce. *Becca Forward, Becca Frontward*. New York: Lothrop, Lee & Shepard Books (1986).

McMillan, Bruce. *Eating Fractions*. New York: Scholastic (1994).

Mathews, Louise. *Gator Pie*. New York: Dodd, Mead & Co. (1979).

Morozumi, Atsuko. *One Gorilla*. New York: Farrar, Straus & Giroux (1990).

Myllar, Rolf. *How Big Is A Foot?* New York: Dell Publishing (1991).

Pienkowski, Jan. *Numbers*. New York: Little Simon (1981).

Pienkowski, Jan. *Sizes*. New York: Little Simon (1990).

Pluckrose, Henry. *Capacity*. New York: Franklin Watts (1988).

Pluckrose, Henry. *Length*. New York: Franklin Watts (1988).

Pluckrose, Henry. *Numbers*. New York: Franklin Watts (1988).

Pluckrose, Henry. *Weight*. New York: Franklin Watts (1988).

Sendak, Maurice. *Chicken Soup with Rice: A Book of Months*. New York: Scholastic (1986).

Sendak, Maurice. *One Was Johnny*. New York: Harper & Row Publishers (1991).

Sitomer, Mindel, and Sitomer, Harry. *Circles*. New York: Thomas Y. Crowell Publishers (1971).

Sitomer, Mindel, and Sitomer, Harry. *How Did Numbers Begin?* New York: Thomas Y. Crowell Publishers (1976).

Spier, Peter. *Fast Slow, High Low: A Book of Opposites*. New York: Doubleday & Co (1972).

Testa, Fulvio. *If You Look Around You*. New York: Dial Books for Young Readers (1983).

Trinca, Rod, and Argent, Kerry. *One Wooly Wombat*. New York: Kane/Miller Book Publishers (1985).

Tudor, Tasha. *One is One*. New York: Macmillan Publishing Company (1956).

Wildsmith, Brian. *Brian Wildsmith's 1, 2, 3's*. New York: Franklin Watts (1987).

APPENDIX C
PROMPTS FOR WRITING IN MATHEMATICS

What I learned in math today.

The hardest thing I did in math was. . . .

I like math because. . . .

Write a letter to a student in a grade below you to tell them about math in the present grade.

What I like about math.

What I don't like about math.

Pictorial Journal—with words, pictures, and numbers.

What I really want to know about math is. . . .

I can solve math problems when. . . .

Math is easy if. . . .

The trouble with math is. . . .

I often make mistakes on math tests when. . . .

My first experience in math was. . . .

What did you learn as you solved the problem?

What questions would you like to ask?

Create an original story problem for a classmate or partner.

Present a graph or a table. Have students write a story that describes or represents the data on the graph or table.

Display a picture. Have students crate a problem about the picture that can be solved mathematically.

Cut pictures from magazines. Have students work in groups to create mathematical story problems about the picture.

List 4 topics or concepts from a chapter you are studying. Have students write a summary describing one of the topics or concepts.

Brainstorm (with students) a list of concepts studied. List concepts on chalkboard. Work in pairs, small groups, or as a whole class. Classify the terms by common elements. Then have pairs or groups produce a paragraph describing one of the categories.

Write for 5 minutes everything you know about such concepts as:

- Circle
- Square
- Fraction
- Decimal
- Length
- Operation

Write a math autobiography. Have students detail their experiences with math.

Discuss mathematicians of "yore." Write a letter to one of the mathematicians discussed.

Write original stories involving mathematical ideas.

Write a play involving math.

Rewrite a page in the textbook.

Write self-help directions describing how a problem was solved.

Describe thinking while solving a problem.

Record questions, predictions, opinions, and categories for a math lesson.

Word Web—Key concept or phrase in center. Have students respond with knowledge or feelings they have about the concept or phrase.

Interview someone in another grade about mathematics. Have a group create the questions to be used in the interview. Record answers during the interview. Discuss and compare activities completed.

Write an article for a school newspaper describing a mathematics activity or activities completed in your class.

Write "original" definitions of selected math concepts or terms.

Keep a log containing notes taken in math class. (This is a verbal explanation of the student's perception of what he or she understands.)

Describe how to complete an operation—step-by-step.

Write about specific help you gave or received from your group during a math lesson.

Examine a problem that is incorrectly solved. Write an explanation about what is incorrect and one justifying your solution.

Use the writing process to create a math story and solution.

- Prewriting—brainstorm, pictures, menus, maps, newspapers.
- Writing—teacher asks questions, challenge.
- Conference—teacher/peers/feedback.
- Revision.
- Publications—on poster, on file cards, or bind.

Read a story. Students write a parallel story.

Read and revise a fairy tale or folk tale to include numerical information. Then generate an original story problem to accompany the tale.

Write about what you did in math, what you learned, and questions you have. Include something you learned or you're not sure about, or something you're wondering about.

Write about what was easy and what was difficult for you in solving the problem.

Write a poem about one or more mathematical concepts.

APPENDIX D
BIBLIOGRAPHY FOR ORAL COMMUNICATION

Ball, Deborah L. "What's all this talk about disclosure?" *Arithmetic Teacher* (November, 1991), pp. 44–48.

Baroody, Arthur J. *Problem Solving, Reasoning, and Communications, K–8; Helping Children Think Mathematically*. New York: Macmillian Publishing Co. (1993).

Carnegie Council on Adolescent Development. *Turning Points: Preparing American Youth for the Twenty-First Century*. Washington, DC: Carnegie Corp. (1989).

Clark, Sally. "Math the write way." *Teaching K–8* (January, 1992), pp. 64–66.

Greenes, Carole, Schulman, Linda, and Spungin, Rika. "Stimulating communication in mathematics." *Arithmetic Teacher* (October, 1992), pp. 78–83.

Kamii, Constance. *Young Children Reinvent Arithmetic*. New York: Columbia Teachers College Press (1986).

Katterns, Bob, and Carr, Ken. "Talking with young children about multiplication." *Arithmetic Teacher* (March, 1984), p. 14.

Lappan, Glenda, and Schram, Paula W. "Communication and reasoning: Critical dimensions of sense making in mathematics." In: *New Directions for Elementary School Mathematics. 1989 Yearbook of the NCTM* (1989), pp. 13–30.

Lietzinger, Larry P. "Problem solving: Tips for teachers: Spotlight on techniques—the four step method of problem solving." *Arithmetic Teacher* (September, 1985), pp. 38–39.

Matz, Karl A., and Leisr, Cynthia A. "Word problems and the language connection." *Arithmetic Teacher* (April, 1992), pp. 18–22.

McKeachie, Wilbert J. *Teaching Tips*. Lexington, MA: D.C. Heath & Co. (1984).

Mumme, Judith, and Shepard, Nancy. "Communication in mathematics." *Arithmetic Teacher* (September, 1990), pp. 18–22.

National Council of Teachers of Mathematics. *Curriculum and Evaluation Standards for School Mathematics*. Reston, VA: National Council of Teachers of Mathematics (1989).

National Council of Teachers of Mathematics. *Professional Standards for School Mathematics*. Reston, VA: National Council of Teachers of Mathematics (1991).

Payne, Joseph, editor. *Mathematics for the Young Child*. Reston, VA: National Council of Teachers of Mathematics (1990).

Small, Marian, Cranahan, Richard, and Ramberg, Thomas. *Data from Classroom Observations for Topic S-2*. Madison, WI: Research and Development Center for Individualizing Schooling (1980).

Stigler, James W. "The use of verbal explanation in japanese and american classrooms." *Arithmetic Teacher* (October, 1988), pp. 27–29.

Timpson, William M., Schlelein, Marty, Jones, Barbara A., and Stephens-Carter, Sherri. *The Naive Expert Challenges the Limits of Young Learners*. (May, 1990).